Contemporary Topics in

POLYMER SCIENCE

Volume 1
Macromolecular Science
Retrospect and Prospect

Contemporary Topics in
POLYMER SCIENCE

Volume 1
Macromolecular Science
Retrospect and Prospect

Edited by
R.D. Ulrich
General Electric Plastics Division
Noryl Products Department
Selkirk, New York

PLENUM PRESS · NEW YORK AND LONDON

Library of Congress Cataloging in Publication Data

Main entry under title:

Macromolecular science—retrospect and prospect.

(Contemporary topics in polymer science; v. 1)
Includes index.
1. Macromolecules—Addresses, essays, lectures. 2. American Chemical
Society. Division of Polymer Chemistry. I. Ulrich, R. D. II. Series.
QD381.7.M3   547'.842   78-6116
ISBN 978-1-4684-2855-1   ISBN 978-1-4684-2853-7 (eBook)
DOI 10.1007/978-1-4684-2853-7

Proceedings commemorating the Silver Anniversary of
the Division of Polymer Chemistry, Inc. of
The American Chemical Society, 1976

©1978 Plenum Press, New York
Softcover reprint of the hardcover 1st edition 1978
A Division of Plenum Publishing Corporation
227 West 17th Street, New York, N.Y. 10011

Contributors

F. A. Bovey
Polymer Chemistry
Research Department
Bell Laboratories
Murray Hill, New Jersey 07974

F. R. Eirich
Polytechnic Institute of New York
333 Jay Street
Brooklyn, New York 11201

John D. Ferry
Department of Chemistry
University of Wisconsin
Madison, Wisconsin 53706

Paul J. Flory
Department of Chemistry
Stanford University
Stanford, California 94305

Maurice L. Huggins
135 Northridge Lane
Woodside, California 94062

Leo Mandelkern
Department of Chemistry
Institute of Molecular Biophysics
Florida State University
Tallahassee, Florida 32306

H. F. Mark
Dean Emeritus
Polytechnic Institute of New York
333 Jay Street
Brooklyn, New York 11201

C. S. Marvel
Department of Chemistry
University of Arizona
Tucson, Arizona 85721

Bryce Maxwell
Polymer Materials Laboratory
Department of Chemical Engineering
Princeton University
Princeton, New Jersey 08540

Herbert Morawetz
Department of Chemistry
Polytechnic Institute of New York
333 Jay Street
Brooklyn, New York 11201

Maurice Morton
Institute of Polymer Science
University of Akron
Akron, Ohio 44325

Charles G. Overberger
Vice President for Research
University of Michigan
Ann Arbor, Michigan 48104

A. Peterlin
Polymers Division
U.S. Department of Commerce
National Bureau of Standards
Washington, D. C. 20234

C. E. Schildknecht
Department of Chemistry
Gettysburg College
Gettysburg, Pennsylvania 17325

J. K. Stille
Department of Chemistry
Colorado State University
Fort Collins, Colorado 80523

Michael Szwarc
State University of New York
Polymer Research Center
College of Environment Science
  and Forestry
Syracuse, New York 13210

Robert D. Ulrich
General Electric Plastics Division
Noryl Products Section
Selkirk, New York 12158

Speakers at the 25 Year Anniversary of the ACS Division of Polymer
Chemistry, Atlantic City, New Jersey, September, 1974. Back row,
left to right: R.F. Boyer, J.C.H. Hwa, F.H. Winslow, R.D. Ulrich,
J.K. Stille, F.R. Mayo; front row, left to right: J.R. Elliot,
C.S. Marvel, O. Vogl, H.F. Mark, P.O. Powers.

Foreword

In 1974, as we approached the National Bicentennial and the Centenary of the American Chemical Society, Professor Otto Vogl, then Chairman of the Division of Polymer Chemistry, arranged a very special symposium dedicated to a review of the history of the Division. It was an extraordinary occasion which included remarks by Professors Herman Mark, Charles Marvel, William Bailey, and Charles Overberger, all past Chairmen of the Division.

The Executive Committee of the Division of Polymer Chemistry felt that 1976 deserved even more attention since it was to be also the 25th, the Silver Anniversary, of the Division of Polymer Chemistry. This year would be a most appropriate one not only to review milestones in our discipline, but also to look to the future.

It was decided to undertake this volume and Dr. R. D. Ulrich agreed to serve as editor in assembling the collected papers. It is the hope of the officers of the Division that this volume will serve many purposes - a reference text, a record, and a source of perspective.

F. E. Bailey
Chairman (1976)
Division of Polymer Chemistry
American Chemical Society

Contents

The History of ACS Division of Polymer Chemistry, Inc.

R.D. Ulrich

Prior to 1946 there was no one division in the American Chemical Society which dealt solely with polymers. Papers in the field were spread among several divisions with a large number concentrated in the Division of Paint, Varnish and Plastics Chemistry.

Interest in the field of polymer chemistry became so intense that a High Polymer Forum, similar to our present symposium system, was established. The wide range of interest in the polymer field was evidenced by the several divisions sponsoring the Forum: Paint, Varnish and Plastics; Physical, Inorganic; Rubber Chemistry; Cellulose; Colloid; Organic; and Petroleum. From 1946 to 1949, 162 papers were presented at 27 sessions of The High Polymer Forum. The attendance at each meeting averaged over 300 and was often limited by the available facilities. Almost at the inception of the High Polymer Forum, discussion regarding the organization of a Division of High Polymer Chemistry was begun.

The ACS by-laws state that: "Any fifty members of the Society in good standing, wishing to organize a Division of the Society, shall set forth in a signed petition, addressed to the Council, the proposed name of the Division, the particular field of Society interest to be stimulated and developed by the proposed Division, and the reasons why it is deemed wise and expedient to establish the Division...".

A large part of the initiative for the formation at the new Division came from within the Division of Paint, Varnish and Plastics Chemistry, as well as the Industrial and Engineering Chemistry Division, and was spearheaded by individuals such as

1

M.L. Huggins and P.O. Powers. The necessary petition, containing
78 signatures, largely from the New York - New Jersey area, was
submitted to the National ACS in October, 1949.

In September, 1950, E.H. Volwiler, then president of the ACS,
announced formation of the Division of High Polymer Chemistry
(later shortened to the Division of Polymer Chemistry). C.S. Marvel,
noted for his work with synthetic rubber during W.W. II and at that
time associated with the University of Illinois, was appointed
chairman. H.F. Mark, internationally recognized as a leading
polymer scientist and director of The Polymer Research Institute
at the Polytechnic Institute of Brooklyn, was appointed secretary-
treasurer. Other officers included W.E. Hanford, R.F. Boyer,
R.M. Fuoss, J.B. Nichols, P.J. Flory, F.R. Mayo and L.A. Wood - a
distinguished list.

Concurrent with the formation of the new division, the first
meeting was held in Chicago with Topics of Free Radicals in Polymerization, Polymer Solutions and Reactions of Macromolecules being
treated in the first symposia. One year later, with a membership
of 414 and dues of $2.00, full divisional status was granted. As
stated in by-law, Art. 1, Sec. 2., the purpose of the Division is,
"to advance knowledge and understanding of the processes of polymerization and of the chemical constitution and chemical and physical properties of polymeric materials and to promote basic
research in these fields."

Almost 25 years have passed since the inception of this
Division. During this time, polymers have gained tremendously in
both their scientific and commercial importance. Has the Division
kept pace with this growth? Has it been responsive to the needs
of both the scientific community and society as a whole? It is
hoped that the following short review of the Polymer Division's
history will provide affirmative answers to these questions.

The most obvious growth comparison to be made is that of the
polymer industry to the membership in the Division. Figure 1
shows the pounds of plastics and resins produced each year in the
U.S. from 1939 to 1972. Notice that the data are in billions of
pounds. The growth has been almost exponential in nature. If
usage of polymeric materials versus such materials as steel were
plotted on a volume basis, the results would be even more striking.
It should also be noticed that polymer production did not really
begin its growth until after W.W. II, about the time the Polymer
Division was coming into creation.

Figure 2 depicts the growth of membership in the Division
from 1950 to 1973. Comparison with the previous production
figures shows an approximately parallel relationship. However,

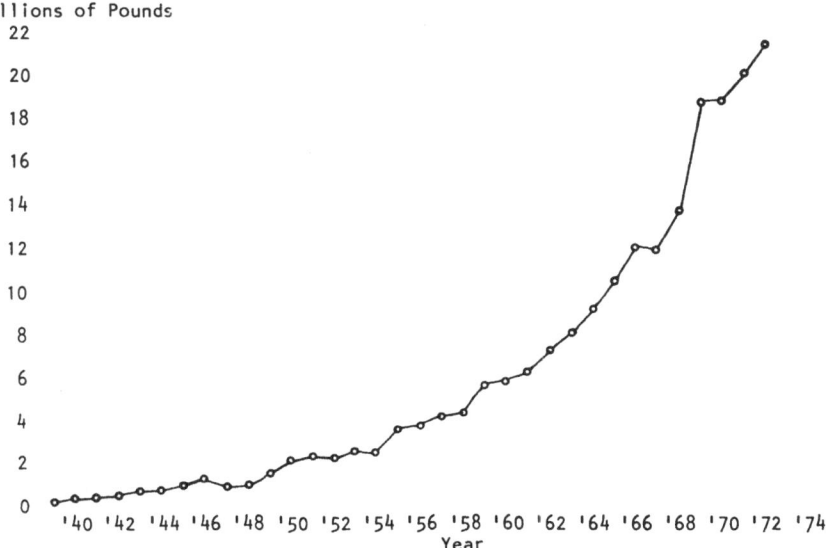

Figure 1.   Resin and Plastics Production

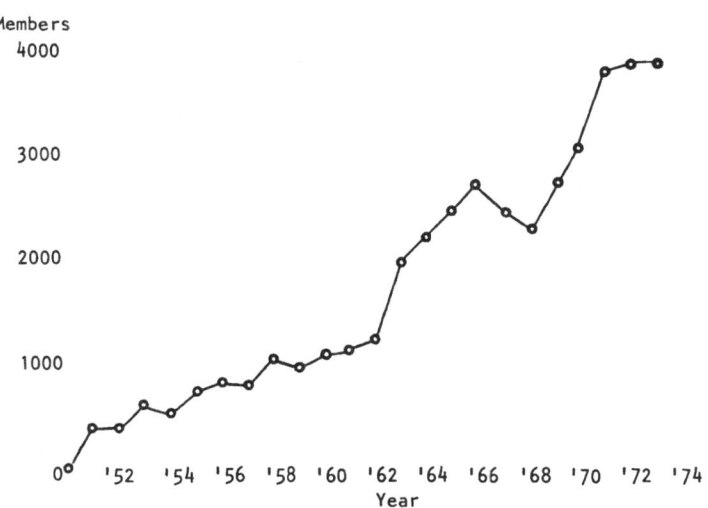

Figure 2.   Growth of Polymer Division Memberships

there is much more here than a cause and effect relationship.
The contributions and hard work of the members and officiers of
the Polymer Division must be considered. This will be pointed out
as the development of polymer industry, education and the Division
are compared.

The appearance of the polymer industry as an important segment
of this country's economy occurred shortly after W.W. II. Prior to
this time, production was limited by shortages caused by the war
effort as well as the newness of the materials themselves. About
the only polymers available for production in the early 1950's were
nylons, polyethylene, polystyrene, phenolics, cellulosics, PVC,
acrylics, silicones, ABS, polyesters, and rubbers. Today there
are approximately 40 major classes of polymeric materials produced
in commercial quantities as well as some rather exotic materials
developed for specific uses. During this 25 year period, new polymers, new methods of producing already important polymers as well
as new ways of utilizing these materials, were being developed.

Some of the most important materials developed and used in
large amounts during this time period were epoxies, fluoropolymers,
polycarbonates, polyethers, polypropylene, and polyurethanes. Of
course, the areas of free-radical, cationic and anionic polymerization grew concurrently. Perhaps the two most important developments in the area of synthesis were those of heterogeneous and
stereoregular polymerization.

It has been suggested that the past 25 years of polymer history can be subdivided into three overlapping decades. As already
pointed out, the 1940's and 1950's saw the major research thrust
in the area of synthesis. The physics of polymeric materials was
vigorously studied in the 1950's and 1960's with such areas as
crystallization, rheology, thermal analysis, and configurational
statistics receiving attention.

The 1960's and 1970's appear to be a decade of engineering
applications. Areas receiving considerable support at this time
are composites and medical applications.

Of course, there are exceptions to this chronological scheme
such as the study of solution properties. Polymer scientists are
also quite cognizant of the fact that polymers can rarely be usefully considered from the standpoint of their chemistry or physics
alone. An understanding of structure-property relationships demand
consideration of the chemistry, physics and engineering application
of the polymer in question.

The prior classification scheme should also **not be construed**
to mean that research and development in each area has ceased.

Perhaps it would be more accurate to describe each period as that when interest in each area accelerated at a goodly rate and plateaued. What lies ahead cannot be predicted, but the tremendous use of polymers insures their continued technical importance.

Has the Polymer Division managed to stay current with major trends in research? Table 1 lists, by year, the various symposia topics presented. If one considers the developments of the last two and one half decades, and when they occured, it becomes obvious that the Division has presented material in new fields of interest as well as those of continuing importance.

Education in the polymer field has grown at a rate that is as impressive as that for production of the materials themselves. At the time this division began, there was only one school with a specific program in polymers: the Polytechnic Institute of Brooklyn with its Polymer Research Institute. Polymer studies at other institutions were largely instances of isolated efforts by dedicated individuals in the classic area of organic chemistry. This was soon to change, however.

At the present time, there are probably three additional institutions with major influence in polymer academia.

The Polymer Science and Engineering Program at the University of Massachusetts consists of a balanced program in the areas of polymer physics, chemistry and engineering. This group was recently distinguished by its designation as an NSF Center of Excellence in Materials Science. This is a milestone for a polymer program and a recognition of legitimacy for the polymer field of which all its researchers can be proud.

Case Western Reserve University with its Division of Macromolecular Science deals mostly with the physics of polymeric materials. Case Western has provided a basis for industrial growth through its training of polymer scientists, a large number of whom join duPont, an acknowledge leader in polymer research and production.

The Department of Polymer Science at the University of Akron has been another group contributing to polymer education. Their strength in the area of rubbers is demonstrated by the large number of their graduates involved with the tire industry.

Today, approximately 1840 credits in polymer science and engineering are available at slightly over 100 institutions in the United States. Included on this list are at least 32 institutions offering programs or degrees in the field. Table 2 lists these.

Recognition of the growth of polymer industry and education over the last 25 years obviates the reason for the parallel growth of the Polymer Division. The same individuals and institutions contributing to the growth of the field were at the same time contributing to the development of the Division. If one were to check a list of officers, committee members, symposia chairmen, and contributors of papers, a parallel to the outstanding institutions and individuals in the field could quickly be drawn. For example, duPont has been the employer of more symposia chairmen than any other single institution.

Up to this point, the growth and history of the Polymer Division has been treated in rather general terms. A consideration of specifics is now in order.

Certain individuals stand out as major contributors to the growth of the Division. A list of officers and committee chairmen are given in Table 3.

During the early years of the Polymer Division, H.F. Mark served over a 5 year period as secretary-treasurer, vice chairman and finally as chairman. His efforts laid a firm foundation for the Division.

The record for continuous service must go to F.H. Winslow. Dr. Winslow served 6 consecutive years as secretary-treasurer, during which Polymer Preprints was begun in 1960, a term as vice-chairman, and one as chairman. It was also during his service in 1961 that the first IUPAC International Symposium on Macromolecular Chemistry was held. After this, he served as a Councilor and chairman and member of various committees. This is a record of very active service lasting from 1956 to the present. The Division owes this man a large debt for his energies and the sense of continuity he has given us over the years.

The division underwent tremendous membership growth from 1962 to 1967, becoming the second largest ACS division in 1964. It was during this time that W.J. Bailey served in the various officers positions. Notable accomplishments during Dr. Bailey's service were the formation of the Education Committee in 1963, from which the Speakers Bureau and Several ACS short courses emanated and the holding of the first Biennial Polymer Symposium in 1962.

E.M. Fettes served as a committee chairman and in various officer's positions from 1964 to 1972. While Dr. Fettes served as vice-chairman in 1965, we became the first ACS division to sponsor a charter flight to Europe for an IUPAC meeting. These flights to Europe have continued. During his term as chairman in 1966, a charter flight to Japan was undertaken. These charter flights have continued with one to South America this year.

In 1966, we gained our first international members. This undertaking was continued by Otto Vogl in 1968 with a membership drive in Europe and Japan. The success of his international efforts is evidenced by the fact that approximately 12% of our members are located outside the U.S.

1968 also saw the beginning of the ACS publication Macromolecules with F.H. Winslow as the editor. And, while on the topic of publications, Polymer Preprints should be mentioned. The Preprints has the largest circulation (equal to The Division membership plus several hundred for institutions and non-members), and lowest price for its size, of any journal in the polymer field. Indeed, the cost and size of the Polymer Preprints has become the limiting factor with regard to the number of papers that can be accepted at our meetings.

J.R. Elliot and J.C.H. Hwa have officially served the Division since 1966 and 1967 respectively. From the beginning of their service, these men have worked energetically for the modernization of the Polymer Division. These efforts were culminated in 1969 when Dr. Hwa, serving as secretary, initiated the official re-organization of the Division to its present business-like form. This restructuring is apparent from Table 3. The Division owes J.C.H. Hwa its gratitude for his foresight and leadership.

1970 saw NSF list Macromolecular Chemistry as a separate heading in the specialties list of The National Register of Scientific and Technical Personnel. Yet another indication of the importance of polymer science.

The axiom that history repeats itself appears to also hold true for the Polymer Division. It will be recalled that we were conceived from the efforts of several divisions with overlapping interests. In 1971, through the efforts of O. Vogl and J.R. Elliot, the Secretariat for Macromolecules, consisting of the Polymer; Organic, Coatings, and Plastics; Rubber; Colloid and Surface; and Wood, Cellulose and Fiber Division was formed. The purpose of this body is to enhance co-operation and coverage of macromolecular subjects in the Division programs. The first Symposium of the Macromolecular Secretariat was held at the 1973 Chicago meeting with "Acrylonitrile in Polymers" as the subject matter.

The Polymer Division has also been responsive to social issues. For example, in 1971 it was stated as policy that, "To keep polymer chemistry relevant to social and environmental prob-lems of our times, the interaction of polymer science with these problems should be highlighted, especially via symposia". In this regard, symposia dealing with environmental protection, housing, waste disposal, population control, water desalination and medical uses have been and will be presented.

It should also be pointed out that our Division was the first to take action regarding the professional problems of our members with formation of the Committee on Professional Matters in 1972.

Incorporation of our Division in 1972 was a logical extension of Dr. Hwa's efforts. The third article of incorporation best sums up the philosophy of the Polymer Division:

"The corporation is organized for the following purposes: The purposes of the corporation are exclusively scientific, educational, charitable and no other, and in furtherance of only these purposes the objects of the corporation shall be within the scope of the science of chemistry as applied to polymers, to encourage in the broadest and most liberal manner the advancement of chemistry and the understanding of the processes of polymerization and of the chemical constitution and chemical and physical properties of the polymeric materials; the promotion of research in chemical science and industry; the improvement of the qualifications and usefullness of chemists through high standards of professional ethics, education, and attainments; the increase and diffusion of chemical knowledge; and by its meetings, professional contacts, reports, papers, discussions, and publications, to promote scientific interests and inquiry, thereby fostering public welfare and education, aiding the development of our country's industries, and adding to the material prosperity and happiness of our people."

Our current officers, headed by Chairman Otto Vogl, are continually updating the Polymer Division to keep it relevant and responsive to the needs of its members and society as a whole, while maintaining its leadership among the various ACS divisions.

The history of the Polymer Division has been one of responsive, positive growth and accomplishment. There is still room for growth. It has been estimated that there are 10,000 to 20,000 potential members for this Division, but in order to reach our potential, we must employ the efforts of all our members. Signing up new members is not enough. The quality of this Division must also continue to grow. If nothing else has been pointed out by this short history, it is that the Polymer Division's status and growth is the direct result of the efforts of its members.

TABLE 1

SYMPOSIA TOPICS

1951

Inorganic Polymers
New Methods and Products in
  Polycondensation
Second Stage Crosslinking

1956

Water Soluble Polymers Other Than
  Cellulose
Fluorine Containing Polymers
New Olefin Polymers

1957

Application of Polymer Science to
  Biochemical Problems

1958

The Crystalline State in Polymers
Inorganic and Other High Tempera-
  ture Polymers
Instruction in Polymer Chemistry
Polyethers and Condensation
  Polymers

1959

Ionic Polymerization
Problems in Polymer Chemistry

1960

Solution Properties of Polymers
Recent Advances in Free Radical
  Polymerization
Microstructure of Proteins
Novel Polymerization
Problems in Polymerization
Absorption of Macromolecules

1961

Structure & Properties of Net-
  work Polymers
Monomers & Polymers from
  Petroleum
Copolymers of Alpha Olefins
Flow Behavior of Polymer Melts
  & Solutions

1962

Characterization of Stereospecific Polymers
Crystallization of Macro-
  molecules

1963

Electrical Properties of
  Polymers
Stabilization of Polymers
Novel Polymer Structures
Polymer Mixtures and Compata-
  bility
Thermal Analysis of High Poly-
  mers

1964

Polymers Containing Aromatic
  Structures
Surface & Bulk Properties of
  Polymers
Stereospecific Polymerization

1965

Recent Developments in Polymer
  Science
Low Angle Scattering from Fib-
  rous & Partially Oriented
  Systems
Transport Phenomena in Polymeric
  Films

TABLE 1
(continued)

1966

New Concepts in Anionic Polymeri-
zation
Meaning of Crystallinity in Poly-
mers
Polymerization & Structure of
Poly-aldehydes
Carbohydrates in Synthetic Poly-
mers
New Concepts in Cationic Polymer-
ization
New Concepts in Emulsion Polymers
Fluorine Containing Polymers
Electro- and Electrooptic Prop-
erties of Macromolecules
Fiber Spinning and Drawing

1967

Application of Computers to Poly-
mer Problems
Ladder and Spiro Polymers
Chemical Reactions of Polymers
Competitive Kinetics in Polymer
Chemistry
Micro Techniques in Polymer Eval-
uation
Mechanism of Mechanical Failure
in Polymers
Polymer Spectroscopy
Polymer Reinforced Plastics
Analytical Gel Permeation Chroma-
tography
Chemical Education with Respect
to Polymer and Coatings Chemis-
try
Polymerization in the Solid State

1968

Properties of Solid Ionic Poly-
mers
Photodegradation of Polymers
Physical Properties of Polymer
Crystals

1968 - continued

Statistical Mechanics of Macro-
molecules
Flammability and Combustion of
Polymeric Compositions
Reactivity of Polymer Solutions
Analytical Calorimetry
Ions and Ion Pairs in Non-
Solvolytic Organic Reactions
Fibers of Thermally Resistant
Organic Polymers
Chemistry of Polyurethanes
Water Vapor Transport in Poly-
mers
Polymer Nonenclature
Radiation Polymerization of
Polymers

1969

Photopolymers
Rheology of Polymers
New Step-Growth Polymerization
Reactions
Engineering Plastics
Polymer Cations
Block Copolymers
Plastic Deformation of Polymers
Permselective Membranes

1970

Multicomponent Polymer Systems
Polymerization - Depolymeriza-
tion Equilibria
Copolymers of - Olefins
Medical Applications of Plastics
Scanning Electron Microscopy
Kinetics and Mechanism of Stereo-
regular Polymerization
Glass Transition Phenomena in
Plastics and Coatings
Properties of Polymers in Inter-
faces
Characterization of Copolymers

TABLE 1
(continued)

1970 - continued

Recent Advances in Free Radical
  Initiators
Highly Cross-linked Network Poly-
  mers
Colloidal and Morphological
  Behavior of Block and Graft
  Copolymers
Small Angle X-ray Scattering of
  Polymers
Determination of Molecular Weight
  of Polymers
Colloidal Polymers

1971

Symposium on the Occasion of the
  Centennial of Rayleight Scatter-
  ing Theory
Mechanically Induced Reactions of
  Polymers
Organometallic Chemistry - Organo-
  metallic Catalysis
Unsolved Problems in Polymer
  Science
Water Desalination
Recent Trends in Determination of
  Molecular Weight
Metal - Organic Plastics Systems
New Addition and Condensation
  Polymers
Dielectric Properties of Polymers
Characterization and Properties of
  Branched Polymers
Recent Advances in Organofluorine
  Chemistry - Organofluorine Poly-
  mers
Thermal and Oxidative Properties
  of Plastics

1972

Use of Lasers in Polymer Science
Polymerization of Cyclic Ethers
  and Sulfides

1972 - continued

Polymerization and Polyconden-
  sation Processes
High-Modulus Wholly Aromatic
  Fibers
The Polymer Chemistry of
  Membranes
Polymers as Chemical Reagents
Synthesis and Characterization
  of Graft Copolymers
Novel Instrumentation in Poly-
  mer Science
Polymerization and Related
  Reactions by Meathesis
N-Chelated Li-Compounds,
  Structure and Uses
Inorganic Polymers
Learning Chemistry from Macro-
  molecules
Non-Polluting Coatings and
  Coating Processes
Transport Through Polymer Films
Polymers and Ecological Problems
Water Soluble Polymers
Thermomechanical Analysis of
  Polymer Solids

1973

Synthesis, Characterization and
  Properties of Highly Alterna-
  ting
Polymer Precursors of Carbon
Recent Advances in Crosslinking
  of Polymers
Polymer Characterization of
  Thermal Methods of Analysis
Configurational Characteristics
  of Chain Molecules
Rheological Properties and Molec-
  ular Structure of Polymers
Reinforced Polymers
Chemistry and Rheology of PVC

TABLE 1
(continued)

1973 - continued

Symposium on Smoke and Toxic Gas
    Evolution from Burning Mater-
    ials
Recent Advances in Polymer Blends
Polymer Impregnated Solids
Ion Containing Polymers
Chemistry and Physics of Polymer
    Powders
Science and Technology of Poly-
    mers of Acrylonitrile

1974

New Look in Co-ordination Poly-
    merization
Structure vs. Mechanical Behavior
    of Polymers
Affinity Chromatography
Polymer - Biological Agent Grafts
Multicomponent Polymer Systems
Polyethers
Physical Structure of the Amor-
    phous State
Properties of Concentrated Poly-
    mer Solutions
History of Polymers
Preparation and Uses of Oligomers
Polymeric Membranes for Separa-
    tion and Purification
New Developments in Additives for
    Coatings and Polymers

1975

Plasma Chemistry of Polymers
Emulsion Polymers; Synthesis,
    Properties and Application
Polymerization and Polycondensa-
    tion Processes
Science and Technology of Adhe-
    sion or Adhesives
New Fluorine Containing Polymers
Hydrogel Grafts for Biomedical
    Applications

1975 - continued

Photoactive Polymers: Chemistry
    and Physics
Properties of Anisotropic Bulk
    Polymers
Reactions at Polymer Surfaces
Stress Induced Crystallization

1976

Milestones in Polymer Chemistry
Synthesis and Properties of
    Condensation Polymers, Espec-
    ially Aromatic Polyamides
Polymers in Health Concerns
New Polymers and Polymerization
Deformation Yield and Fracture
    of Polymers
Controlled Release Polymeric
    Formulations
Learning Chemistry from Macro-
    molecules
Macromolecules and Societal
    Needs
Polymer Reagents
Structure-Solubility Relation-
    ship in Polymers
Charge Transfer Copolymerization
Chemical Spectroscopy
Relaxation Phenomena in Polymers
Water-Borne Polymers

1977

Stabilization and Degradation
    of Polymers
Rigid Chain Polymers: Synthesis
    and Properties
Structure and Properties of
    Block Copolymers
Polymeric Drugs
New Ring Opening Polymerizations
Toughened Polystyrene
Advances in Organo-Metallic
    Polymers

<u>TABLE 1</u>
(continued)

<u>1977</u> - continued

Novel Instrumental Methods for
  Probing Polymer Structures
Chromatography of Polymers
Crystallinity Control and Polymer
  Mechanics
Polymerization in Mesophases and
  Mesomorphic Order in Polymers

Table 2

Programs and Degrees in Polymer Science and Engineering

Institution	Total Credits	Organizational Unit	Degrees Offered
1. University of Akron	93	Department of Polymer Science	M.S. and Ph.D. in Polymer Science
2. Bronx Community College	28	Department of Chemistry, Chemical Technology and Plastics Technology	A.A.S. Degree in Chemical Technology-Plastics Technology Option
3. Case Western Reserve University	59	Division of Macromolecular Science	B.S. in Engineering-Polymer Science Major; M.S. and Ph.D. in Macromolecular Science and Engineering
4. Columbia University	9	Polymer Program, Division of Applied Chemistry	Departmental Degrees
5. University of Connecticut	33	Polymer Science Program, Institute of Materials Science	M.S. and Ph.D. in Materials Science
6. Cornell University	21	Materials Science Center	Degrees by participating departments
7. Ferris State College	22	Industrial Department	Associate Degree in Plastics Technology
8. University of Florida	16	Center for Macromolecular	Degrees by participating departments
9. Lehigh University	15	Materials Research Center	Degrees by participating departments
10. Lowell Technological Institute	120	Department of Plastics Technology; Polymer Science Program, Department of Chemistry	B.S. in Plastics Technology, M.S. in Polymer Science and Ph.D. in Chemistry-Polymer Science Option

Table 2 (Cont.)

Institution	Total Credits	Organizational Unit	Degrees Offered
11. University of Maryland	27	Applied Polymer Science Program, Departments of Chemical Engineering and Chemistry	M.S. and Ph.D. in Applied Polymer Science
12. University of Massachusetts	58	Polymer Science and Engineering Progtam	M.S. and Ph.D. in Polymer Science and Engineering
13. Massachusetts Institute of Technology	24	Polymer Materials Program	Degrees by participating departments
14. University of Michigan	21	Macromolecular Research Center	M.S. and Ph.D. in Macromolecular Science and participating departments
15. Milwaukee School of Engineering	17	Plastics Technology Program, Engineering School	Certificate of a Plastic Technologist
16. University of Missouri, Rolla	12	Materials Research Center	Degrees by participating departments
17. Montclair State College	15	Plastics Technology Program, Department of Industrial Education and Technology	Departmental Degree
18. Newark College of Engineering	33	Division of Technology, Evening School	Certificate Program in Plastics
19. State University of New York College of Forestry at Syracuse University	17	State University Polymer Research Center	B.S. in Forest Chemistry, Wood and Polymer Chemistry Option
20. North Carolina State University	29	School of Textiles	Ph.D. in Fiber and Polymer Science
21. North Dakota State University	39	Department of Polymers and Coatings	B.S., M.S., and Ph.D. in Chemistry-Polymers and Coatings Major

Table 2 (Cont.)

Institution	Total Credits	Organizational Unit	Degrees Offered
22. Northwestern University	16	Materials Research Center	Degrees by participating departments
23. Pennsylvania State University	27	Polymer Science Section, Department of Materials Science	B.S. in Polymer Science
24. Polytechnic Institute of Brooklyn	32	Polymer Research Institute, Department of Chemistry; Department of Chemical Engineering	M.S. in Polymeric Materials; Ph.D. in Chemistry or Chemical Engineering
25. Princeton University	27	Polymer Science and Materials Program	Degrees by participating departments
26. Rensselaer Polytechnic Institute	31	Polymer Science and Engineering Program	Degrees by participating departments
27. Richmond College, CUNY	30	Department of Chemistry	M.A. and Ph.D. in Polymer Chemistry
28. University of Southern Mississippi	91	Department of Polymer Science	B.S. in Plastics-Technology; B.S., M.S. and Ph.D. in Polymer Science
29. Stevens Institute of Technology	29	Department of Chemistry and Chemical Engineering; Plastics Institute of America	Departmental degrees
30. Texas A&M University	21	Polymer Research Center, Department of Chemistry	Departmental degrees
31. University of Utah	41	Division of Materials Science and Engineering	Divisional degrees
32. Washington University	30	Materials Science and Engineering Program	Degrees by participating departments

Table 3
ACS POLYMER DIVISION OFFICERS AND COMMITTEE CHAIRMEN

	1951	1952	1953
Chairman	C.S. Marvel	W.E. Hanford	P.J. Flory
Vice-Chairman	W.E. Hanford	P.J. Flory	R.M. Fuoss
Secretary	H.F. Mark	H.F. Mark	H.F. Mark
Treasurer	H.F. Mark	H.F. Mark	H.F. Mark
Councillors		R.F. Boyer	R.F. Boyer
		W. Stockmayer	W. Stockmayer
Alternate Councillors		H.M. Spurlin	H.M. Spurlin
		P.M. Doty	P.M. Doty
Program		H.F. Mark	B.H. Zimm
Nominating	W.O. Kenyon	W.O. Kenyon	J.B. Nichols
Education			
Biennial Symposium			
Program			
Arrangements			
Membership		J. Dec	J. Dec
Publications - Preprints			
Nomenclature			
By-laws			
Charter Flights			
Local Group			
Public Relations			
Professional Matters			
International			
Membership Booth			
Planning			
Circulation			
Awards			
Grants			

Table 3
(continued)

	1954	1955
Chairman	R.M. Fuoss	H.F. Mark
Vice-Chairman	H.F. Mark	R.F. Boyer
Secretary	J. Dec	J. Dec
Treasurer	J. Dec	J. Dec
Councillors	R.F. Boyer	J.D. Ferry
	J.D. Ferry	T.G. Fox
Alternate Councillors	C.E. Schlidknecht	C.E. Schildknecht
	P.M. Doty	A.V. Tobolsky
Program	M.T. O'Shaughnessy	A.V. Tobolsky
Nominating	J.H. Nichols	C. Walling
Education		
Biennial Symposium		
Program		
Arrangements		
Membership	W.F. Busse	R.T. Dean
Publications - Preprints		
Nomenclature		
By-laws		
Charter Flights		
Local Group		
Public Relations		
Professional Matters		
International		
Membership Booth		
Planning		
Circulation		
Awards		
Grants		

Table 3
(continued)

	1956	1957
Chairman	R.F. Boyer	T.G. Fox
Vice-Chairman	T. G. Fox	A.V. Tobolsky
Secretary	J. Dec	F.H. Winslow
Treasurer	J. Dec	F.H. Winslow
Councillors	J.D. Ferry	T.G. Fox
	T.G. Fox	A.V. Tobolsky
Alternate Councillors	C.E. Schildknecht	J.D. Ferry
	A.V. Tobolsky	F.H. Winslow
Program	S.G. Weissberg	S.G. Weissberg
Nominating	E.M. Fettes	A.M. Borders
Education		
Biennial Symposium		
Program		
Arrangements		
Membership		
Publications - Preprints		
Nomenclature		
By-laws		
Charter Flights		
Local Group		
Public Relations		
Professional Matters		
International		
Membership Booth		
Planning		
Circulation		
Awards		
Grants		

Table 3
(continued)

	1958	1959
Chairman	A.V. Tobolsky	F.R. Mayo
Vice-Chairman	F.R. Mayo	C.G. Overberger
Secretary	F.H. Winslow	F.H. Winslow
Treasurer	F.H. Winslow	F.H. Winslow
Councillors	A.V. Tobolsky	A.V. Tobolsky
	F.R. Mayo	F.R. Mayo
Alternate Councillors	J.D. Ferry	J. Dec
	F.H. Winslow	F.H. Winslow
Program	E.M. Fettes	C.G. Overberger
Nominating	J. Dec	T.G. Fox
Education		
Biennial Symposium		
Program		
Arrangements		
Membership		
Publications - Preprints		
Nomenclature		
By-laws		
Charter Flights		
Local Group		
Public Relations		
Professional Matters		
International		
Membership Booth		
Planning		
Circulation		
Awards		
Grants		

Table 3
(continued)

	1960	1961
Chairman	C.G. Overberger	T. Alfrey, Jr.
Vice-Chairman	T. Alfrey, Jr.	M. Morton
Secretary	F.H. Winslow	F.H. Winslow
Treasurer	F.H. Winslow	F.H. Winslow
Councillors	F.R. Mayo	T. Alfrey, Jr.
	T. Alfrey, Jr.	M. Morton
Alternate Councillors	F.H. Winslow	F.H. Winslow
	C.K. Rosenbaum	C.K. Rosenbaum
Program	T. Alfrey, Jr.	M. Morton
Nominating	F.R. Mayo	W.F. Busse
Education		
Biennial Symposium		
Program		
Arrangements		
Membership		
Publications - Preprints	C.K. Rosenbaum	C.K. Rosenbaum
Nomenclature		
By-laws		
Charter Flights		
Local Group		
Public Relations		
Professional Matters		
International		
Membership Booth		
Planning		
Circulation		
Awards		
Grants		

Table 3
(continued)

	1962	1963
Chairman	M. Morton	A.M. Bueche
Vice-Chairman	A.M. Bueche	F.H. Winslow
Secretary	F.H. Winslow	W.J. Bailey
Treasurer	F.H. Winslow	W.J. Bailey
Councillors	T, Alfrey, Jr.	F.H. Winslow
	M. Morton	M. Morton
Alternate Councillors	F.H. Winslow	W.J. Bailey
	C.K. Rosenbaum	C.K. Rosenbaum
Program	A.M. Bueche	F.H. Winslow
Nominating	W.E. Gibbs	L.A. Wall
Education		
Biennial Symposium		
Program	H. Starkweather	
Arrangements	J.B. Kinsinger	
Membership		
Publications - Preprints	C.K. Rosenbaum	C.K. Rosenbaum
Nomenclature		
By-laws		
Charter Flights		
Local Group		
Public Relations		
Professional Matters		
International		
Membership Booth		
Planning		
Circulation		
Awards		
Grants		

Table 3
(continued)

	1964	1965
Chairman	F.H. Winslow	R. Simha
Vice-Chairman	R. Simha	E.M. Fettes
Secretary	W.J. Bailey	W.J. Bailey
Treasurer	W.J. Bailey	W.J. Bailey
Councillors	R. Simha	R. Simha
	F.H. Winslow	F.H. Winslow
Alternate Councillors	C.K. Rosenbaum	C.K. Rosenbaum
	E.M. Fettes	E.M. Fettes
Program	R. Simha	E.M. Fettes
Nominating	L.H. Dunlap	S.G. Weissberg
Education		
Biennial Symposium		
Program	F.R. Bovey	
Arrangements	W.R. Krigbaum	
Membership	E.M. Fettes	J. Lal
Publications - Preprints	C.K. Rosenbaum	C.K. Rosenbaum
Nomenclature		
By-laws		
Charter Flights		
Local Group		
Public Relations		
Professional Matters		
International		
Membership Booth		
Planning		
Circulation		
Awards		
Grants		

Table 3
(continued)

	1966	1967
Chairman	E.M. Fettes	W.J. Bailey
Vice-Chairman	W.J. Bailey	W.H. Stockmayer
Secretary	W.E. Cass	W.E. Cass
Treasurer	J.B. Kinsinger	J.B. Kinsinger
Councillors	R. Simha	E.M. Fettes
	E.M. Fettes	W.H. Stockmayer
Alternate Councillors	C.K. Rosenbaum	W.E. Cass
	W.E. Cass	J.B. Kinsinger
Program	W.J. Bailey	W.H. Stockmayer
Nominating	F.H. Winslow	G.B. Butler
Education	M. Morton	M. Morton
Biennial Symposium	J.R. Ell	J.R. Elliott
Program	J.R. Elliot	J.K. Stille
Arrangements	E. Baer	R.S. Porter
Membership	J. Lal	J.C.H. Hwa
Publications - Preprints	C.K. Rosenbaum	R.W. Lenz
Nomenclature		R.B. Fox
By-laws		F.H. Winslow
Charter Flights		H.Z. Friedlander
Local Group		
Public Relations		
Professional Matters		
International		
Membership Booth		
Planning		
Circulation		
Awards		
Grants		

Table 3
(continued)

	1968	1969
Chairman	W.H. Stockmayer	J.R. Elliot
Vice-Chairman	J.R. Elliot	J.B. Kinsinger
Secretary	W.E. Cass	J.C.H. Hwa
Treasurer	J.B. Kinsinger	R.S. Porter
Councillors	E.M. Fettes	E.M. Fettes
	W.H. Stockmayer	J.B. Kinsinger
Alternate Councillors	W.E. Cass	R.S. Porter
	J.B. Kinsinger	J.C.H. Hwa
Program	J.R. Elliot	J.B. Kinsinger
Nominating	R.F. Boyer	O. Vogl
Education	M. Morton	M. Morton
Biennial Symposium		J.K. Stille
Program		T.G. Fox
Arrangement		
Membership	J.C.H. Hwa	B.M. Culbertson
Publications - Preprints	R.W. Lenz	R.W. Lenz
Nomenclature	R.B. Fox	H.K. Livingston
By-laws	F.H. Winslow	F.H. Winslow
Charter Flights	H.Z. Friedlander	H.Z. Friedlander
Local Group	E. Perry	E. Perry
Public Relations		
Professional Matters		
International		
Membership Booth		
Planning		
Circulation		
Awards		
Grants		

Table 3
(continued)

	1970	1971
Chairman	J.B. Kinsinger	W.E. Gibbs
Vice-Chairman	W.E. Gibbs	J.P. Kennedy
Secretary	J.C.H. Hwa	J.C.H. Hwa
Treasurer	O. Vogl	O. Vogl
Councillors	E.M. Fettes	J.B. Kinsinger
		E.M. Fettes
Alternate Councillors	R.S. Porter	R.S. Porter
	J.C.H. Hwa	J.C.H. Hwa
Program	W.E. Gibbs	J.P. Kennedy
Nominating	D.M. McIntyre	R. Van Deusen
Education	E.M. Pearce	E.M. Pearce
Biennial Symposium	J.K. Stille	J.K. Stille
Program	T.G. Fox	
Arrangements	H.L. Stephens	J.E. Mark
Membership	B.M. Culbertson	A.R. Cain
Publications - Preprints	R.W. Lenz	H.L. Stephens
Nomenclature	H.K. Livingston	N.M. Bikales
By-laws	L.H. Kunlap	L.H. Kunlap
Charter Flights	H.Z. Friedlander	C.A. Garber
Local Group		
Public Relations	J.F. Pendleton	J.F. Pendleton
Professional Matters		
International	J.P. Kennedy	V.T. Stannett
Membership Booth		
Planning		
Circulation	R. Saxon	R. Saxon
Awards		
Grants		

Table 3
(continued)

	1972	1973
Chairman	J.P. Kennedy	J.C.H. Hwa
Vice-Chairman	J.C.H. Hwa	O. Vogl
Secretary	F.E. Karasz	F.E. Karasz
Treasurer	O. Vogl	J.E. Mark
Councillors	J.B. Kinsinger	J.B. Kinsinger
	E.M. Fettes	W. Gibbs
Alternate Councillors		
Program	J.C.H. Hwa	O. Vogl
Nominating	J.R. Elliot	J.F. Pendleton
Education	E.M. Pearce	E.M. Pearce
Biennial Symposium Program Arrangements	J.K. Stille	J.K. Stille
Membership	A.R. Cain	J.F. Pendleton
Publications - Preprints	H.L. Stephens	H.F. Stephens
Nomenclature	N.M. Bikales	L.G. Donaruma
By-laws	L.H. Dunlap	L.H. Dunlap
Charter Flights	C.A. Garber	C.A. Garber
Local Group		
Public Relations	J.F. Pendleton	J.C. Salamone
Professional Matters	A.E. Pavlath	A.E. Pavlath
International	V.T. Stannett	V.T. Stannett
Membership Booth		J.F. Kinstle
Planning		J.C.H. Hwa
Circulation	R. Saxon	R. Saxon
Awards		J.P. Kennedy
Grants		J.L. Koenig

Table 3
(continued)

	1974	1975
Chairman	O. Vogl	J.K. Stille
Vice-Chairman	J.K. Stille	F.E. Bailey
Secretary	F.E. Karasz	E.J. Vanderberg
Treasurer	J.E. Mark	J.C. Salamone
Councillors	J.B. Kinsinger	W.D. Gibbs
	W. Gibbs	J.C.H. Hwa
Alternate Councillors		B.M. Culbertson
		R.W. Lenz
Program	J.K. Stille	B.M. Culbertson
Nominating	W.C. Wooten	R.T. Conley
Education	E.M. Pearce	E.M. Pearce
Biennial Symposium	J.K. Stille	J.R. Schaefgen
Program	F.H. Winslow	E.M. Pearce
Arrangements	J.C. Salamone	
Membership	J.F. Pendleton	J.F. Pendleton
Preprints Editor	J.E. McGrath	
Nomenclature	L.G. Donaruma	L.G. Donaruma
By-Laws	L.H. Dunlap	L.H. Dunlap
Charter Flights	H.Z. Friedlander	H.Z. Friedlander
Local Group		
Public Relations	J.C. Salamone	S.E.B. Petrie
Professional Matters	A.E. Pavlath	A.E. Pavlath
International	V.T. Stannett	V.T. Stannett
Membership Booth	J.F. Kinstle	J.F. Kinstle
Planning	J.C.H. Hwa	E.M. Fettes
Circulation	R. Saxon	R. Saxon
Awards	J.P. Kennedy	P.W. Morgan
Grants	J.L. Koenig	
Intersociety Relations		O. Vogl
Macromolecular Secretariat		J. Elliot
Past Chairman		
Member-At-Large		

Table 3
(continued)

	1976	1977
Chairman	F.E. Bailey	V.T. Stannett
Vice-Chairman	V.T. Stannett	B.M. Culbertson
Secretary	E.J. Vanderberg	E.J. Vanderberg
Treasurer	J.C. Salamone	J.C. Salamone
Councillors	W.D. Gibbs	W.D. Gibbs
	J.C.H. Hwa	J.C.H. Hwa
Alternate Councillors	R.W. Lenz	R.W. Lenz
	P.W. Morgan	P.W. Morgan
Program	B.M. Culbertson	F. Harris
Nominating	J. Economy	J.P. Kennedy
Education	E.M. Pearce	E.M. Pearce
Biennial Symposium	J.R. Schaefgen	M. Shen
Program	E.M. Pearce	H. Hopfendberg
Arrangements		S.C. Israel
Membership	A.E. Pavlath	A.E. Pavlath
Preprints Editor	R.M. Ikeda	R.M. Ikeda
Nomenclature	L.G. Donaruma	L.G. Donaruma
By-laws	L.H. Dunlap	L.H. Dunlap
Charter Flights	H.Z. Friedlander	H.Z. Friedlander
Local Group		
Public Relations	S.E.B. Petrie	S.E.B. Petrie
Professional Matters	A.E. Pavlath	A.E. Pavlath
International	H.J. Harwood	H.J. Harwood
Membership Booth	J.F. Kinstle	J.E. McGrath
Planning	E.M. Fettes	E.M. Fettes
Circulation	R. Saxon	R. Saxon
Awards	P.W. Morgan	W.C. Wooten
Grants		
Intersociety Relations	O. Vogl	O. Vogl
Macromolecular Secretariat	F.E. Bailey	F.E. Bailey
Past Chairman	J.K. Stille	F.E. Bailey
Member-At-Large	P.H. Lindenmeyer	P.H. Lindenmeyer

Table 3
(continued)

<u>1978</u>

Chairman   B.M. Culbertson
Vice-Chairman   E. Vandenberg
Secretary   J.E. McGrath
Treasurer   J.C. Salamone
Councillors   F.E. Bailey
                                      W.E. Gibbs
                                      J.C.H. Hwa
                                      P.W. Morgan
Alternate Councillors   R.J. Ambrose
                                      P.A. Hiltner
                                      J.B. Kinsinger
                                      S.E.B. Petrie

Program   F.W. Harris
Nominating   J.K. Stille
Education   E.M. Pearce
Biennial Symposium   M. Shen
   Program   F.W. Harris
   Arrangements   H. Hopfenberg
Membership   S.W. Shalaby
Preprints Editor   R.M. Ikeda
Nomenclature   L.G. Donaruma
By-laws   J.F. Kinstle
Charter Flights
Local Group
Public Relations   E. Weissberger
Professional Matters   A.E. Pavlath
International   H.J. Harwood
Membership Booth   J.J. O'Malley
Planning   E.M. Fettes
Circulation   R. Saxon
Awards   W.C. Wooten
Grants   E. Baer
Intersociety Relations   O. Vogl
Macromolecular Secretariat   B. Culbertson
                                      E. Vandenberg
Past Chairman   V. Stannett
Member-At-Large   P. Lindenmeyer

F. A. Bovey

The first studies of polymers by nmr were published only about a year after the first reports of the observation of the NMR phenomenon in bulk matter by Bloch and Purcell in 1946. This early work[1] concerned the solid state, which for the following decade was the sole object of nmr investigations of polymers. By measurements of nuclear relaxation, involving for the most part protons, much important information concerning chain dynamics of a variety of polymers has been gathered.

Another important domain of the application of nmr to polymers concerns the observation and interpretation of their high resolution spectra; here, one is seeking primarily structural rather than dynamic information, and it has been customary to employ solutions of polymers so that the chain motion will be as rapid as possible in order to average out the dipolar broadening effects of direct nuclear interaction. In recent years, however, it has become possible to obtain significant dynamic information from solution spectra and also to obtain high resolution spectra, and therefore structural information, directly from solids. The barrier between two hitherto quite separate domains has thus begun to crumble very significantly. I will touch on these matters again later in the discussion.

In the earlier era of high resolution nmr, when the study of small molecules had reached a fairly advanced state, it was still generally assumed that very large molecules could not give useful spectra even in solution because of their supposedly slow motions, as evidenced by the very high viscosities of such solutions. There was also a feeling that the spectra would be too complex to interpret.

In the late 1950's, investigators finally began to test these
notions, and found that they were in fact unfounded. Reasonably
well resolved spectra of ribonuclease[2] and polystyrene[3] were
reported. It was found that very viscous solutions could give
quite narrow lines. This happy observation was of course a consequence of the fact that, although random coil macromolecules
pervade volumes that are much larger than their molecular volumes
and produce high viscosities through entrapment of solvent molecules and entanglement among themselves, the local motion of the
chain segments may yet remain very fast, i.e. in the nanosecond
or even picosecond range. (For solutions of biopolymers in their
native helical or folded states, however, the dipolar broadening
may indeed be very severe; we shall discuss such systems in Sec.5.)

Stereochemical Configuration

Among the most useful findings was the observation that proton
nmr could be employed, --even with the low frequency spectrometers
of the time--, to obtain quantitative information concerning the
stereochemistry of vinyl polymers. The Ziegler-Natta "revolution"
was then relatively new and was still generating much excitement.
In the classical researches of Natta and his collaborators, X-ray
fiber diagrams were the principal structural tool[4]. This method
required that the polymer be crystallized, which was often difficult or impossible. Despite its great power, X-ray was subject
to errors of interpretation and often failed to reveal significant
structural details.

In Fig. 1 are shown the 40 MHz spectra of chloroform solutions
of two samples of polymethyl methacrylate, as reported by myself
and George Tiers in 1960[5]. Spectrum (a) is that of a polymer prepared with a free radical initiator; spectrum (b) is that of a
polymer prepared with n-butyllithium in toluene, a typical anionic
initiator. The marked differences in the spectra arise from
differences in stereochemical configuration. Let us consider the
chain in the simplest terms which reflect the relative chirality
of adjacent asymmetric α-carbon atoms, i.e. monomer dyads (Fig. 2).
In a racemic or r dyad (a), --the repeating unit of a syndiotactic
chain--, the methylene protons are equivalent in a time average
over the conformations present, even though the planar zigzag conformation shown may not necessarily predominate. They therefore
have the same chemical shift and are expected to appear as singlet.
But in a meso or m dyad, the characteristic unit of an isotactic
chain, the two protons are in non-equivalent environments. This
is obvious for the planar zigzag conformation shown and can be
readily shown to be true for an average over all possible conformations[6,7]. They will therefore in general have different chemical
shifts, and the geminal scalar coupling between them (usually of the
order of 12-16 Hz) will become evident. In polymethyl methacrylate

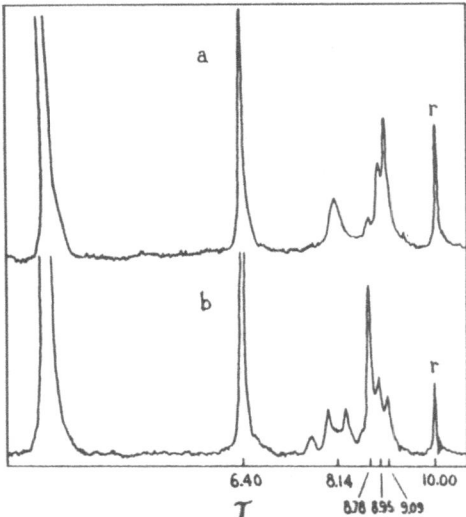

Figure 1. The 40 MHz spectra of PMMA in chloroform. (a) Polymer prepared from free radical initiator; (b) prepared from n-butyl-lithium in toluene, a typical anionic initiator.

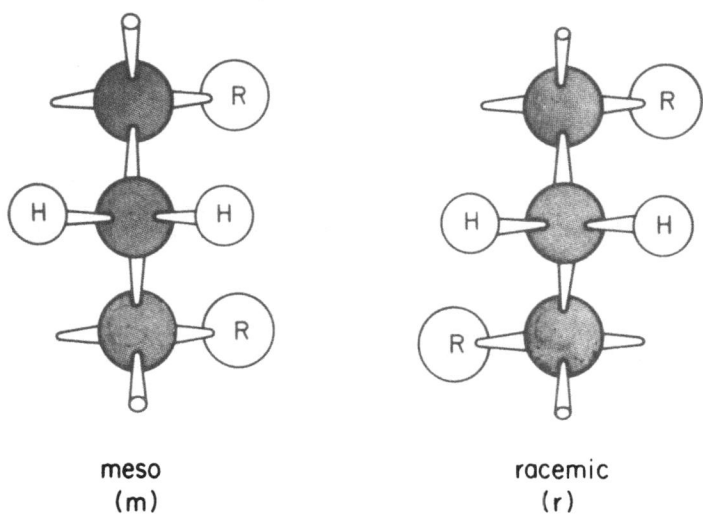

meso
(m)

racemic
(r)

Figure 2. Monomer dyads. (a) Racemic or r dyad; (b) Meso or m dyad.

(and in other α, α´ disubstituted vinyl chains without vicinal
proton-proton coupling) the meso methylene resonance will accordingly be an AB or AX quartet. It is evident that in Fig. 1, the
methylene proton spectrum of the anionically initiated polymer
exhibits such a quartet centered at 8.14τ (1.86δ) and that therefore this polymer is predominantly isotactic. In the free radical
polymer spectrum the methylene protons appear as a singlet, and
so this polymer must be predominantly syndiotactic. Thus, proton
nmr is an absolute method, and no recourse to X-ray is necessary
even if this were possible.

The α-methyl resonances centered at ca. 9τ(1δ) could also be
interpreted to give quantitative estimates of isotactic (abbrevi-
ated mm) and syndiotactic (rr) triad sequences of monomer units,
and also of the mixed unit mr, termed heterotactic, which must
occur in chains which are not perfectly stereoregular:

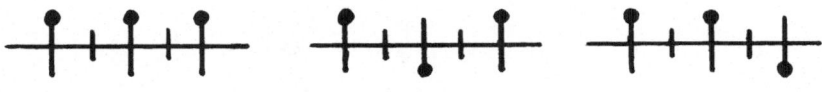

     isotactic triad   syndiotactic triad   heterotactic triad
         (mm)   (rr)   (mr)

(The ester methyl resonance could in principle give the same information but in most solvents, as in chloroform, it is not configura-
tionally sensitive). In Fig. 2, we see three α-methyl proton
resonances having the same chemical shifts but very different intensities in each spectrum. From their measured areas, the triad
probabilities are readily determined.

Such measurements not only demonstrated the stereochemistry
of these polymers, but also enabled one to define certain features
of the mechanism of polymerization. If we assume that the probability of generating a meso sequence when a new monomer unit is
added to the growing chain can be denoted by a single parameter
P_m, then the generation of the chain is a Bernoulli-trial process
with regard to its stereochemistry, the outcome of the next
monomer addition being independent of the stereochemistry of the
chain already formed. The probability of generating a racemic
unit will, of course, be given by $(1-P_m)$. A triad sequence involves two monomer additions; the probabilities of mm, mr, and rr
are P_m^2, $2P_m(1-P_m)$, and $(1-P_m)^2$, respectively. A plot of these
relations is shown in Fig. 3. It will be noted that the proportion of mr units rises to a maximum when $P_m = 0.5$, corresponding
to the generation of a random or atactic polymer; for such a
polymer, the proportion mm:mr:rr will be 1:2:1. For any given

polymer, if Bernoullian, the triad sequence frequences should lie
on a single vertical line in Fig. 3. If this is not the case,
the polymer's configurational sequence is non-Bernoullian and must
be described by two or more conditional probabilities. Free radical
polymers always lie to the left of the center of Fig. 3, and are
(with rare exceptions) Bernoullian within experimental error.
Polymers made with anionic and coordination catalysts are usually
non-Bernoullian. In Fig. 4, all the points to the right of the
center correspond to anionic polymers; the mm triad content has
been arbitrarily placed on the mm curve and the others allowed to
fall where they may. It will be observed that they do not fall on
the Bernoullian curves.

In the years following these relatively primitive observations,
a very large number of vinyl and related polymer systems have been
studied by nmr in many laboratories, and their spectra interpreted
in the way just described. If the resolving power of the spectro-
meter is sufficient, configurational sequences longer than triad
may be observed. For protons, this has generally been accomplished
through the use of instruments operating at high magnetic fields,
up to 83 kilogauss at present, and correspondingly high frequencies;
it is a basic rule in nmr that the separation of chemical shifts
(on a frequency scale) increases in direct proportion to the magnetic field strength. With respect to β-methylene groups one may
expect to resolve tetrad sequences, appearing as a fine structure
on the m and r dyad resonances. As shown in Table 1, the m and r
spectra should be resolved into three tetrad subspectra each, for
a total of six. According to the symmetry properties of the
sequences, the β-methylene protons should be equivalent in two of
these and non-equivalent in the other four, as indicated by the
subscripts a and b in the chain projections in Table 1. One may
also expect α-groups to be resolved into ten different pentad
configurational sequences. If the polymer has been generated by
a Bernoulli trial propagation, the sequence probabilities will be
as indicated in Table 1, and may be plotted in a manner analagous
to Fig. 3[6,7]. Still longer sequences may be observable in some
instances. In general, we will have the following numbers N(n)
of sequences of length n:

n	2	3	4	5	6	7	8 ...
N(n)	2	3	6	10	20	36	72 ..., or

in general $N(n) = 2^{n-2} + 2^{m-1}$

where m=n/2 if n is even and m=(n-1)/2 if n is odd. It is evident
that, regardless of the generating statistics, the number of
possible types of sequences increases rapidly with their length
(asymptotically as 2^{n-2} as $N \to \infty$) and that their discrimination
and assignment beyond hexads will be very difficult.

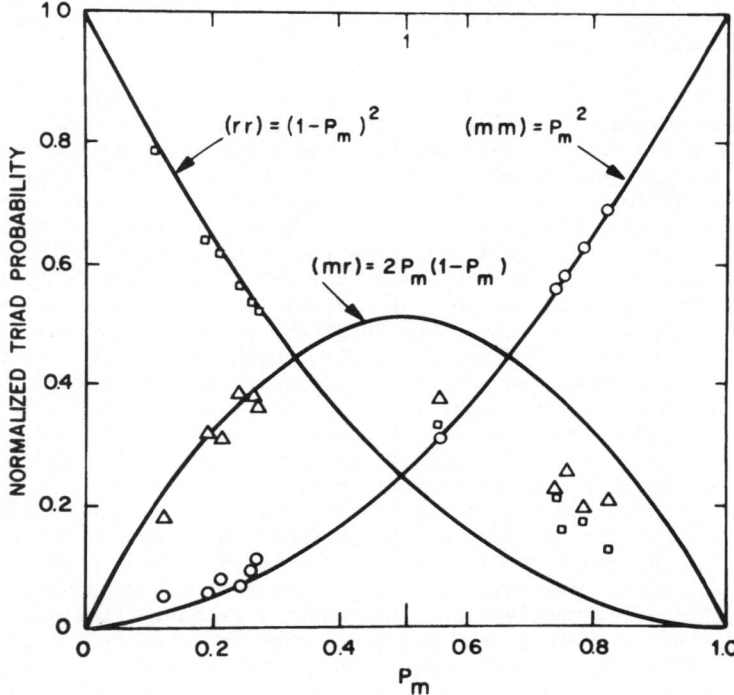

Figure 3. Normalized triad probability vs. Pm, the probability of generating a meso sequence when a new monomer unit is added to the growing chain.

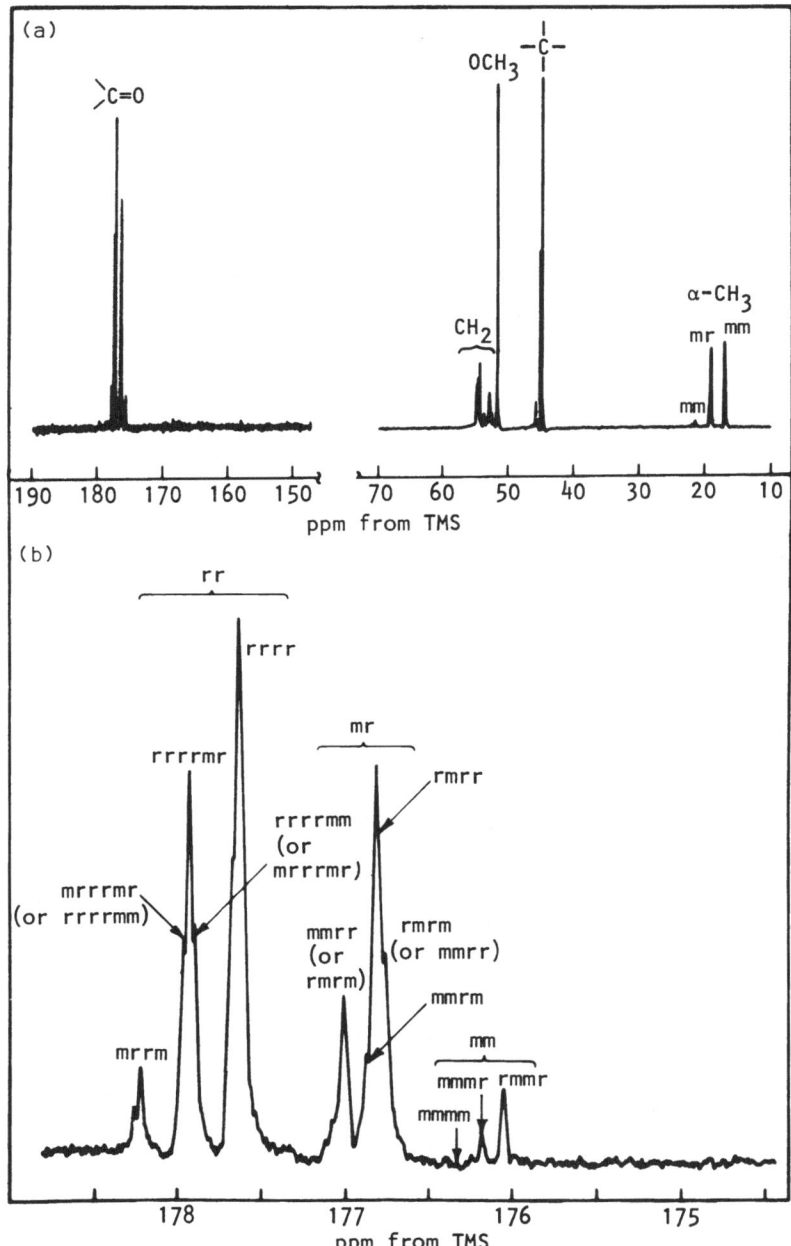

Figure 4. 90 MHZ C_{13} spectrum of free radical PMMA.

Table 1. Possibilities and Sequence Probabilities of Bernoulli Trial Propagation

	α-Substituent			β-CM₂		
	Designation	Projection	Bernoullian probability	Designation	Projection	Bernoullian probability
Triad / Dyad	Isotactic, mm (i)		Pm^2	$meso$, m		Pm
	Heterotactic, mr (h)		$2Pm(1-Pm)$	Racemic, r		$(1-Pm)$
	Syndiotactic, rr (s)		$(1-Pm)^2$			
Pentad / Tetrad	$mmmm$ (isotactic)		Pm^4	mmm		Pm^3
	$mmmr$		$2Pm^3(1-Pm)$	mmr		$2Pm^2(1-Pm)$
	$rmmr$		$Pm^2(1-Pm)^2$	rmr		$Pm(1-Pm)^2$
	$mmrm$		$2Pm^3(1-Pm)$	mrm		$Pm^2(1-Pm)$
	$mmrr$		$2Pm^2(1-Pm)^2$	rrm		$2Pm(1-Pm)^2$
	$rmrm$ (heterotactic)		$2Pm^2(1-Pm)^2$	rrr		$(1-Pm)^3$
	$rmrr$		$2Pm(1-Pm)^3$			
	$mrrm$		$Pm^2(1-Pm)^2$			
	$rrrm$		$2Pm(1-Pm)^3$			
	$rrrr$ (syndiotactic)		$(1-Pm)^4$			

A number of proton studies of polymethyl methacrylate at superconducting frequencies[6],[7] confirm these expectations. Rather than
discussing these, however, let us consider the observation of carbon-
13 spectra. Because of the much greater chemical shift range of
carbon-13 nuclei, --over 250 ppm for structures of interest in
polymers as compared to less than 10 ppm for protons--, carbon
spectra have proved to be of considerably greater power in the
study of the structure of this and many other synthetic polymers,
thanks largely to the development of Fourier transform instruments
with spectrum accumulation. In this technique, no scalar coupling
information is preserved; the protons are normally strongly irradi-
ated to remove their coupling (125-250 Hz) to the carbon-13 nuclei
and collapse the carbon multiplets to singlets. This procedure
increases the effective signal to noise ratio, and also provides
an extra gain in sensitivity owing to the operation of the
nuclear Overhauser effect. This is a carbon-proton cross-relaxation
phenomenon which perturbs the carbon energy levels and enhances
the carbon signal strength by a factor of up to three-fold. (It
may also provide significant information concerning the dynamics
of the chains, as we shall briefly note later.)

In Fig. 4 is shown the 90 MHz C-13 spectrum of free radical
polymethyl methacrylate, the same as that in Fig. 1a. A superconducting instrument was employed, operating at 360 MHz for
protons[9]. Spectrum (a) shows all the resonances; because of the
large (ca. 160 ppm) range of chemical shifts, the spectrum is
necessarily rather compressed. The α-methyl resonance (ca. 17-
22 ppm, referred to tetramethylsilane as zero) is split into
triads with pentad fine structure not discernible on this scale.
The quaternary carbon (ca. 45 ppm) is likewise split, but more
narrowly. The methoxyl carbon, like the methoxyl proton, is
insensitive to configuration. The β-carbons (52-56 ppm) show
tetrad and hexad splitting. Most sensitive is the carbonyl spectrum
at ca. 177 ppm, shown expanded in (b); this shows splitting to the
base line for most of the pentad sequences, and clear indications
of heptad splitting as well; the latter observation indicates
resolution of the resonances of structures differing only in the
relative orientation of carbonyl groups separated by seven inter-
vening carbon-carbon bonds.

With such structural information available complex mechanistic proposals can be tested[8] and rather subtle questions can be
answered. An illustration is furnished by the investigation of
the mechanism of the isotactic polymerization of propylene with
Ziegler-Natta coordination catalysts of the type $TiCl_3 \cdot Al(C_2H_5)_2Cl$
by determining the type of error occasionally introduced.[10],[11]
Such catalysts, which are of course heterogeneous in nature, are
complex in their action. Some sites on the titanium trichloride
crystal surface may produce isotactic chains which are stereoregular to the extent of at least 98%. What is the nature of the

irregularities? Aside from occasional head-to-heat units (rare in
the isotactic polymer), two types may be imagined:

    1. "Template" propagation error:

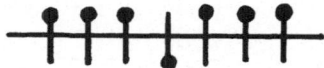

for which the ratio (mmr):(mrr) is 1:1; if pentads can be discriminated, as in the fact they can, one expects to see in
addition to the predominant mmmm sequence, the sequences mmmr,
mmrr, and mrrm in 2:2:1 ratio;

    2. "Steric" propagation error:

for which one will observe no mrr triads and only the mmmr and mmrm
"defect" pentads in a 1:1 ratio.

    The first of these hypothetical possibilities implies that the
propagation is under the control of an asymmetric catalyst site
which is sensitive to the configuration of the growing chain
(formed by monomer insertion at the titanium-carbon bond on the
catalyst surface) and corrects the occasional r propagation error
by restoring the absolute relationship of chain to site. The
second structure implies that the principal correcting force is
the steric nature of the chain-end itself; once a new m unit is
added beyond the r propagation error, propagation proceeds as
before.

    Carbon-13 spectroscopy is particularly effective in the study
of paraffinic polymers because the chemical shifts are very sensitive to structure, being spread over a ca. 45 ppm range as compared only about 2 ppm for protons. In Fig. 5 are shown the ^{13}C
spectra of isotactic, atactic and syndiotactic polypropylenes[12].
Although all three carbons, the CH_2, α-CH, and CH_3, are responsive
to configuration, the CH_3 carbon is clearly the most sensitive.
In Fig. 6, the CH_3 regions of the atactic (a) and the isotactic (b)
polymers are expanded. The resonance assignments shown are critical to the correct identification of the propagation errors, and,
although the point is not beyond dispute[13], appear to have been
securely established by Zambelli et al.[14] by means of model hydrocarbons. The conclusion is that rr sequences are generated and
that the chain irregularity is of the "template" kind, --type[1].
This appears reasonable in view of the generally accepted picture
of the action of coordination catalysts. (The resonance at 19.3
ppm is irrelevant to the problem and may be due to head-to-head
structures.)

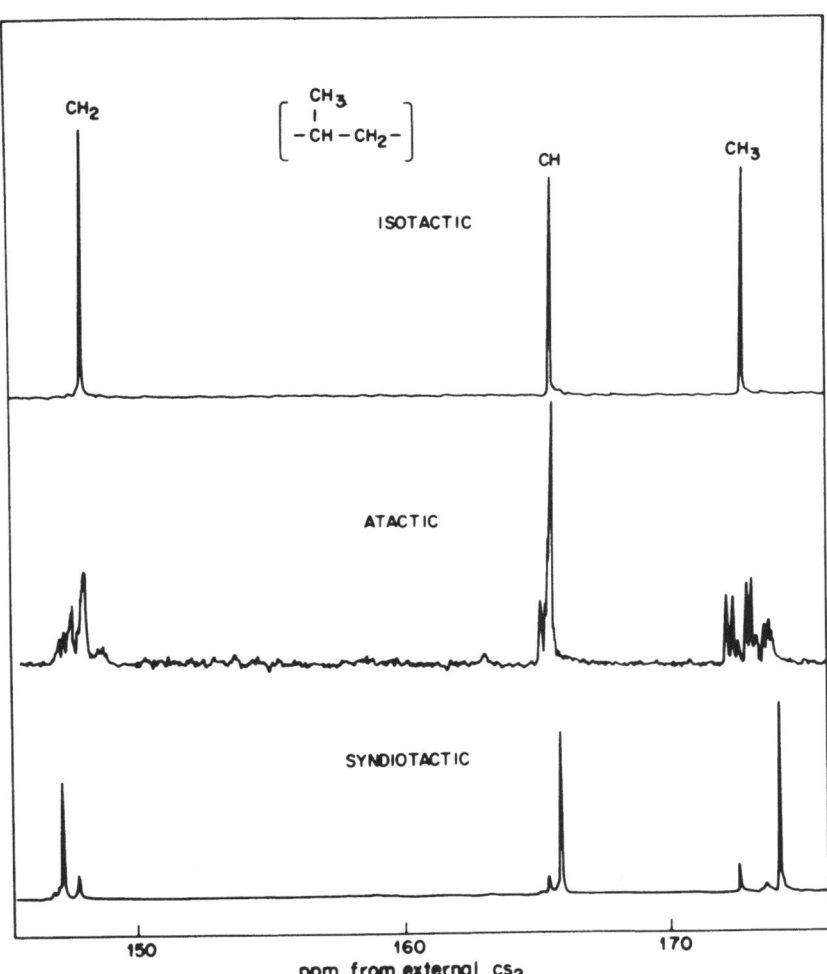

Figure 5. C$_{13}$ spectra of isotactic, atactic and syndiotactic polypropylene.

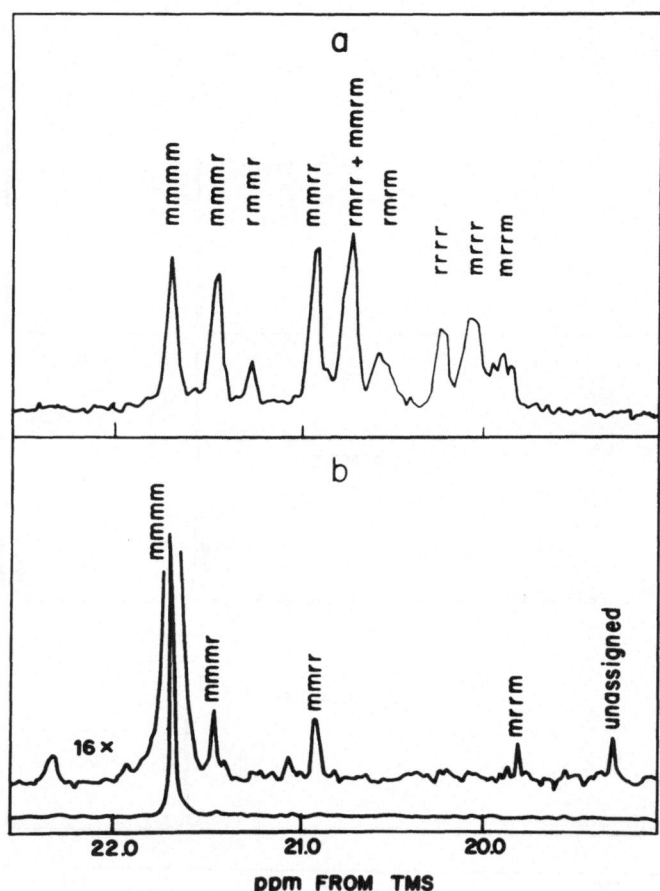

Figure 6.   Expanded CH$_3$ regions of Figure 5.

Copolymer Structure

Another area in which both proton and carbon-13 nmr have proved very powerful is the determination of the structure of copolymers. This has a long history (ref. 6, Chap. X; 15,16), beginning with the observations of butadiene-styrene copolymers in 1959[17]. Again, the information content of the spectra has increased remarkably since these early reports. Although compositional sequence lengths and probabilities can be calculated from the copolymer equation using the traditional data of polymer composition vs. monomer feed composition, nmr allows direct measurement of the sequences and gives in addition much structural detail not available from overall composition alone.

If only compositional sequences are to be considered and discrimination of tetrads is possible, it is evident that as many as ten sequences should be observable:

Dyads	AA	AB(BA)	BB
Triads	AAA		BBB
	BAA(AAB)		ABB(BBA)
	BAB		ABA
Tetrads	AAAA	AABA(ABAA)	BBBB
	AAAB(BAAA)	AABB(BBAA)	BBBA(ABBB)
	BAAB	BABA(ABAB)	ABBA
		BABB(BBAB)	

Generalizing, we find the number $N(n')$ of distinguishable sequences of length n' to be:

n'	2	3	4	5	6	7	8
$N'(n)$	3	6	10	20	36	72	136

the same as the series describing the number of configurational sequences n, where n = n-1. If stereochemical configuration is also to be taken into account, the number of sequences increases very rapidly with n''[5]

n''	2	3	4	5	6
$N(n'')$	6	20	72	272	1056

The analysis of styrene-methyl methacrylate copolymers has been accomplished by Katritzky et al.[18] at the triad level using carbon-13 nmr, and many examples of other copolymers, particularly of dines, have now been reported. A relatively simple system, in

which no stereochemical complications occur, is provided by copolymers of vinylidene chloride and isobutylene, in which neither
monomer generates asymmetric centers. Compositional tetrads are
readily discriminated by proton nmr at 60 MHz.[19,20]. In Fig. 7 is
shown the 360 MHz proton spectrum of a free radical copolymer containing about 65 mol % vinylidene chloride[21]. At least fifty peaks
can be seen and most of them can be assigned without ambiguity
except in the high field region, where some of the isobutylene
resonances are not quite certain. In one region, shown enlarged,
octad sequences can be readily resolved. Probably not even the
most devoted students of copolymer structure and the mechanism of
their formation would care to extract all the statistical information implicit in such spectra, but it shows what can be done.

Branching

Next to molecular weight and its distribution, branching is
the most important structural variable (assuming a given type of
basic chain unit) influencing the properties of polymers and polymer
solutions. It is, of course, well known that short-chain branching
is particularly critical in its effect on the morphology and solid
state properties of polyethylene, while long-chain branching has a
comparably profound effect on solution viscosity and melt rheology.
The traditional method for branch measurement is the observation of
the CH_3 distortion band in the infrared at 1375 cm^{-1}. With proper
correction for band overlap, this procedure can count the branches
correctly, but gives little or no information concerning their
length, type (whether trifunctional or tetrafunctional), or distribution.

Carbon-13 nmr is particularly well adapted to the study of
branching in polyethylene because of the sensitivity of the chemical
shifts to the degree of substitution (already evident in the
spectrum of polypropylene, discussed earlier) and proximity to
chain ends. It is also possible to predict paraffinic carbon chemical shifts for any given structure with some accuracy using a set
of rules first deduced by Grant and Paul[22] from measurements on a
variety of small paraffinic molecules and subsequently refined by
others[23-25]. By use of these rules and by the observation of
appropriate ethylene-olefin copolymers[26-28], it has now been possible to give a fairly complete accounting of the structure of high
pressure polyethylene. In Fig. 8 is shown the 25 MHz C-13 spectrum
of a typical material, together with a computed spectrum based on
chemical shifts taken from ethylene-olefin copolymer spectra[28] and
the branch frequencies (per 1000 CH_2 groups) shown in the lower
right corner, the latter adjusted to match the observed peak intensities. The carbon atom designations are indicated on the right-hand
side of the experimental spectrum and follow the system suggested

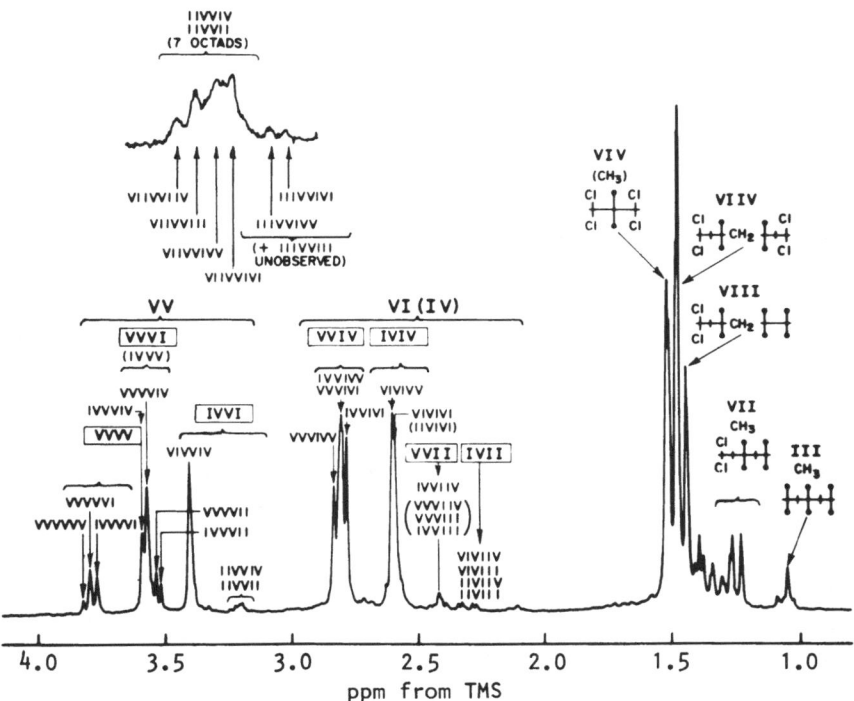

Figure 7. The 360 MHZ proton spectrum of a free radical copolymer
containing ca. 65 mole percent vinylidene chloride.

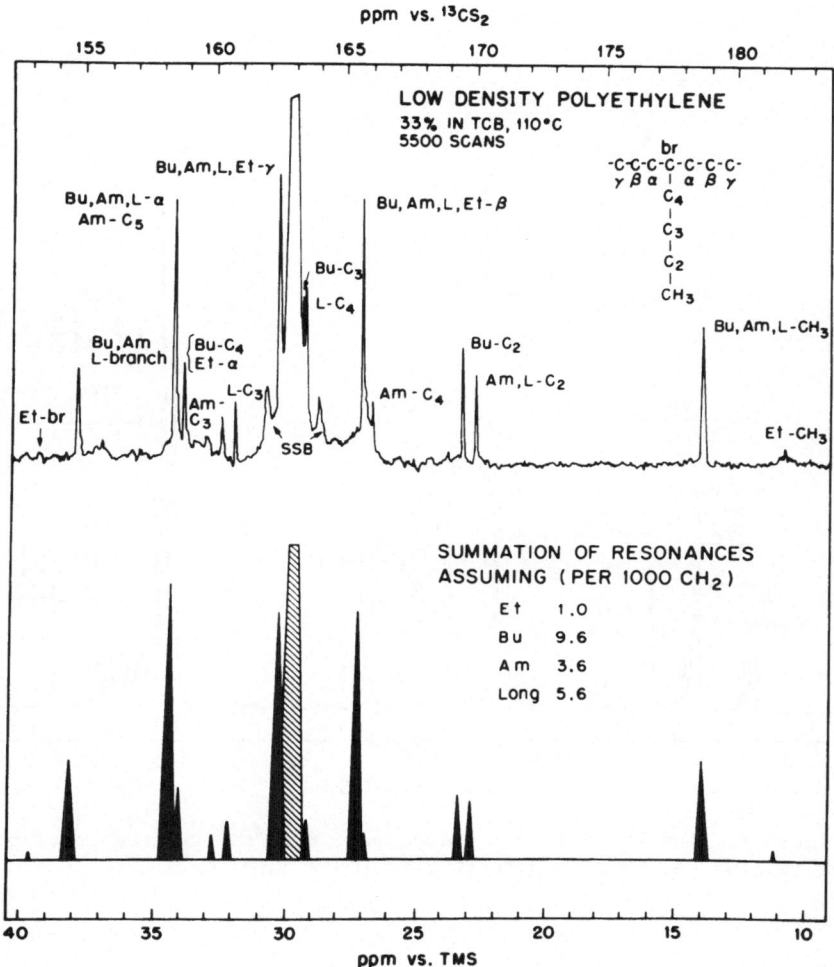

Figure 8.   The 25 MHZ C_{13} spectrum of a low density polyethylene.

by Randall[27]. The principal peak in the spectrum, which is severely
truncated in Fig. 8, is at 30 ppm (vs. tetramethylsilane) and corresponds to those methylene groups which are four or more carbons
removed from a branch point or chain end. The assignments of the
other reseonances are as indicated. It is to be noted that,
although a small number of ethyl branches are detectable, this is
in contrast to the widely accepted conclusion from infrared measurements[29] that such branches are dominant. n-Butyl branches are the
most frequent, followed by n-amyl. There is no trace of methyl or
n-propyl branches. If short branches are generated by the "back-
biting" mechanism suggested many years ago by Roedel[30], the reaction
below

is evidently most probable when n=3 or 4, has a low but finite
probability when n=1, and zero probability when n=0 or 2. All
branches are trifunctional.

Of particular interest is the peak at 32.16 ppm designated
"L-C$_3$". This corresponds to the third carbon from the end of all
branches six or more carbons in length. By careful comparison to
polyethylene fractions characterized by intrinsic viscosity and
light scattering measurements, it has been shown[28] that this resonance in all probability provides a direct measure of truly "long"
branches in polyethylene, i.e. those containing many tends or hundreds of carbons and formed by intermolecular chain transfer. Apart
from its scientific interest, this observation has provided a
measure of a structural variable which in our laboratory has proved
to correlate consistently with the rheological behavior of molten
polyethylenes and with certain important solid state properties as
well.

Biopolymers

A very large and vigorous field of nmr studies is that of biopolymers, including not only biological molecules of natural origin
but model polymers of simpler structure. Actually, the first high
resolution nmr spectrum ever reported (in 1957) was that of a protein,
native ribonuclease[2]. This proton spectrum, observed at 40 MHz, is
shown in Fig. 9. Four broad protein resonances are observable, and
in addition a sharper peak arising from residual HDO. Ribonuclease
is a globular protein of 13700 molecular weight. In its native
state, the polypeptide chains are tightly coiled into a compact
spheroidal structure in which there is little if any main-chain
segmental motion, although certain side chains on the exterior of
the molecule may enjoy some rotational freedom. The braodness of
the peaks is partly due to this cause but also to the fact that at

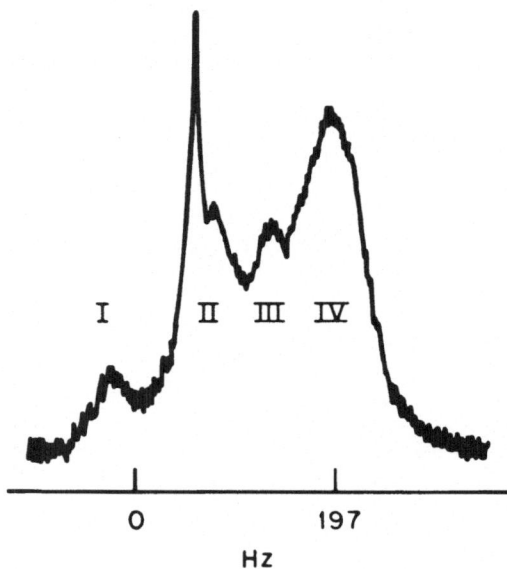

Figure 9. 40 MHZ proton spectrum of native ribonuclease.

40 MHz the many hundreds of chemically shifted lines are merged
into bands. Again, the information content of such spectra has
increased spectacularly as a result of experimental advances over
the last twenty years. In Fig. 10 is shown the 360 MHz spectrum
of carboxypeptidase inhibitor, obtained by Dr. D.J. Patel in this
laboratory[31]. The improvement in resolution is partly due to the
more rapid tumbling of this smaller protein (molecular weight about
4000), but is mainly due to the nearly 10-fold greater dispersion
of chemical shifts in the superconducting field. The resonances at
low field (ca. 7-8δ) arise from aromatic and histidine ring protons.
(Peptide histidine and arginine NH protons would appear at still
lower field, --8-8.5δ--, but have been exchanged for deuterium.)
Then comes a band of resonances at 4-5δ, arising mainly from backbone α-CH protons, followed by a wide variety of side-chain protons
and, at highest field, (0.4-1.5δ) by methyl proton resonances of
valine, leucine, and isoleucine. It will be noted that many of the
resonances show narrow splittings of 6-8 Hz. These correspond to
J-couplings of protons on neighboring carbons, and some of them
contain information concerning the conformations of the side chains.
In H_2O solution, the peptide NH resonances appear, and their splittings, arising from vicinal coupling to the α-CH protons, can provide
information concerning the conformation of the main peptide chain[32].

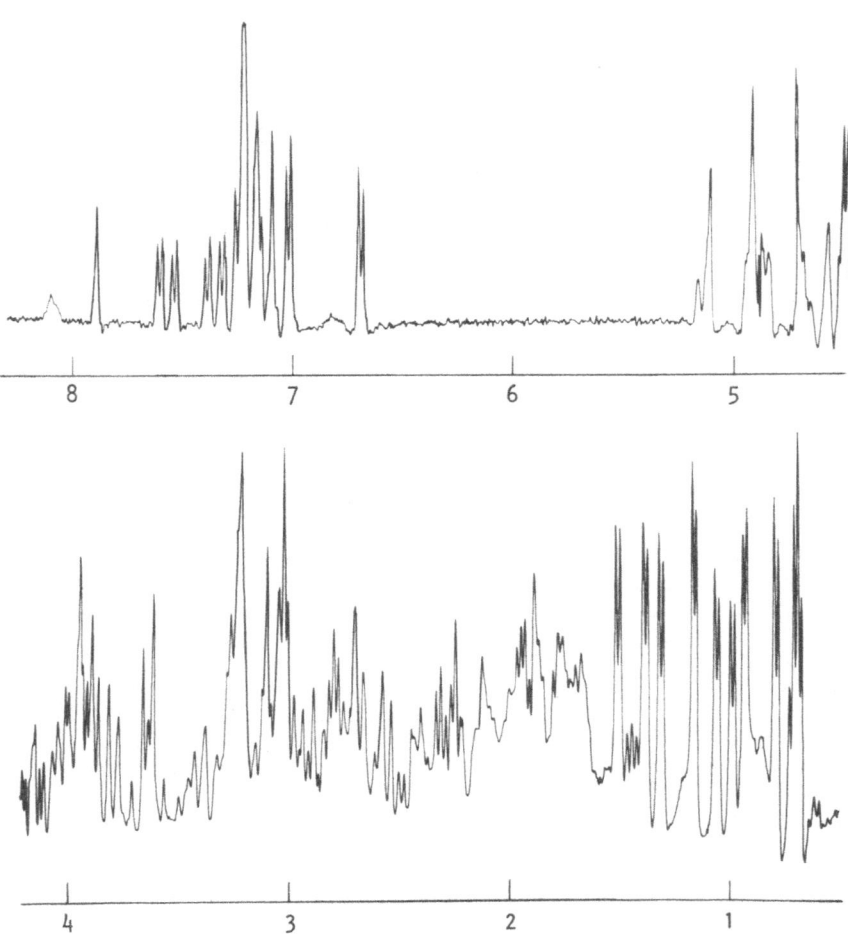

Figure 10. 360 MHZ spectrum of carboxypeptidase inhibitor.

It is not inconceivable that by such coupling measurements, combined
with the observations of the chemical shift effects upon titration
of the ionizable side chains, and possibly by the use of shift
reagents, the conformational structure of this protein (for which
there is at present no X-ray structure although the amino acid sequence is known) could be worked out by nmr alone.

Of the many other aspects of the application of nmr to biopolymers (by not only proton but also ^{13}C and ^{31}P spectroscopy) we
may mention the observation of the much studied α-helix-to-random
coil transition in homo- and copolypeptides[17,32], the study of the
structure of heme proteins, which exhibit remarkable chemical shifts
(ref. 6, Chap. XIV), and the observation of the double helix-random
coil transition in polynucleotides[33].

The Future

High resolution nmr is already in a position to supply more
structural data concerning synthetic macromolecules than can be
readily assimilated and interpreted. No doubt many new polymers
will receive attention in the future and many familiar ones will be
more definitively investigated. It appears at present, however,
that a major effort of the future, --which is indeed already under
way--, will be the use of nmr, principally of ^{13}C, for the study of
the dynamic behavior of polymer chains in the solution and solid
state through the measurement of nuclear spin lattice and spin-spin
relaxation and of the nuclear Overhauser enhancement. The latter
is often found to be less than the maximum three-fold value in
polymer chains, and the implications of this in terms of chain
motion are currently being investigated[34]. Another significant
research area is the use of powerful proton decoupling fields in
pulse experiments (of varying degrees of complexity) to eliminate
both scalar and direct dipole-dipole couplings, and so enable one
to obtain high resolution spectra of polymers in the solid state.
By this means, nmr can be usefully extended to polymers which are
highly cross-linked or otherwise intractable. For soluble polymers,
the structures of which are well known from the solution spectra,
it is very likely that new information concerning chain conformation in the glassy and crystalline states can be obtained since
in these states molecular motion is slow even on the nmr time scale.

The promise of nmr in the study of biological systems is at
least equally great. We may expect it to yield increasingly detailed
information concerning conformations and conformational transitions
of proteins and nucleic acids, the nature of active sites, the
binding of drugs, and the structure and dynamics of membranes and
whole cells.

REFERENCES

1. N. L. Alpert, *Phys. Rev.*, 72, 637 (1947).
2. M. Saunders, A. Wishnia, and J. G. Kirkwood, *J. Amer. Chem. Soc.* 79, 3289 (1957).
3. M. Saunders and A. Wishnia, *Am. N. Y. Acad. Sci.*, 70, 870 (1958).
4. G. Natta, *Chem. Ind.* (Milan) 46, 397 (Nobel Prize Address).
5. F. A. Bovey and G. V. D. Tiers, *J. Polymer Sci.*, 44, 173 (1960).
6. F. A. Bovey, High Resolution NMR of Macromolecules, F. A. Bovey, Academic Press, New York, 1972; Chap III.
7. F. A. Bovey, *Acc. Chem. Res.*, 1, 175 (1968).
8. H. L. Frisch, C. L. Mallows, and F. A. Bovey, *J. Chem. Phys.*, 45, 1565 (1966).
9. F. C. Schilling and F. A. Bovey, unpublished observations.
10. A. Zambelli and C. Tosi, *Adv. in Polymer Sci.*, 15, 32 (1974).
11. A. Zambelli in NMR, Basic Principles and Progress, Vol. 4, p. 101, P. Diehl, E. Fluck, and R. Kosfeld, Eds., Springer-Verlag, Berlin, 1971.
12. A. Zambelli, D. E. Dorman, A. I. R. Brewster, and F. A. Bovey, *Macromolecules*, 6, 925 (1973).
13. J. C. Randall, *J. Polym. Sci.*, Polymer Phys. Ed., 12, 703 (1974).
14. A. Zambelli, P. Locatelli, G. Bajo, and F. A. Bovey, *Macromolecules*, 8, 687 (1975).
15. H. J. Harwood and W. M. Ritchey, *J. Polymer Sci.*, Part B, 3, 419 (1965).
16. H. J. Harwood, *Agew. Chem.*, Int. Ed. Engl. 4, 1051 (1965).
17. F. A. Bovey, G. V. D. Tiers, and G. Filipovich, *J. Polymer Science*, 38, 73 (1959).
18. A. R. Katritzky, A. Smith, and D. E. Weiss, *J. Chem. Soc.* (London) Perkin Trans. II, 1974, 1547.
19. K. H. Hellwege, U. Johnsen, and K. Kolbe, *Kolloid-Z.* 214, 45 (1966).
20. J. B. Kinsinger, T. Fischer, and C. W. Wilson III, *J. Polm. Sci.*, Part B, 4, 379 (1966); 5, 285 (1967).
21. R. E. Cais and F. A. Bovey, unpublished observations.
22. D. M. Grant and E. G. Paul, *J. Amer. Chem. Soc.* 86, 2984 (1964).
23. L. P. Linderman and J. Q. Adams, *Anal. Chem.* 43, 1245 (1971).
24. R. E. Carhart, D. E. Dorman, and J. D. Roberts, unpublished results.
25. C. J. Carman, A. R. Tarpley, Jr., and J. H. Goldstein, *Macromolecules* 6, 719 (1973).
26. D. E. Dorman, E. P. Otocka, and F. A. Bovey, *Macromolecules* 5, 574 (1972).
27. J. C. Randall, *J. Polymer Sci.*, Polym. Phys. Ed., 11, 275 (1973).
28. F. A. Bovey, F. C. Schilling, F. L. McCrackin and H. L. Wagner, *Macromolecules* 9, 76 (1976).
29. A. H. Willbourn, *J. Polymer Sci.* 34, 569 (1959).
30. M. J. Roedel, *J. Amer. Chem. Soc.* 75, 6110 (1953).

31. D. J. Patel, private communication.
32. F. A. Bovey, *Macromol. Rev.* 9, 1 (1974).
33. D. J. Patel and L. Canuel, *Proc. Nat. Acad. Sci.* (USA) 73, 674 (1976).
34. J. Schaefer in Topics in C-13 NMR Spectroscopy, G. Levy, Ed., Wiley-Interscience, New York, 1974.

F. R. Eirich

On arrival at Poly in 1947 from Cambridge, England where Dr. Eirich researched and lectured in the areas of Colloid and Polymer Science, Physical Chemistry, and Rheology, his first contributions were to the physical chemical characterization of polymers in solution, building on the work of Doty and Zimm including the techniques of osmometry, isothermal diffusion, electrophoresis, ultracentrifugation, flow birefringence, and viscosimetry. About ten papers, in cooperation of M. Schick, T. McGoury, J. Riseman, S. Newman, B. Rosen, P. Kamath, R. Aelion and A. Loebl emerged from this research on the relation between particle size, charge and shape and the viscosity and sedimentation of polymer molecules in solution, on the molecular weights of synthetic polypeptides, and on the relation between the coefficients in Einstein's and Huggin's viscosity equations and the concentration correction factors in sedimentation and diffusion.

Also, from '47 to '58, sponsored by DOD, an extended study on polymerization kinetics was undertaken, particularly of allylic compounds, in cooperation with N. Gaylord, N. Field, H. Starkweather, A. Adicoff and M. Litt, rounding out our knowledge of degradative free radical chain transfer and retarded polymerization. This work led to studies of the mechanism of peroxide decomposition as a function of environments, especially of the cage effect, with W. Brown and J. Rajbenbach. This work produced further a series of papers on heat and abrasion resistent organic glasses based on acrylics and allylics, and some of the earliest polymer-polymer laminates. Supplementing this work in the late '60's, a study of the diffusion through polymers over a range of temperatures and pressures was carried out with K. Kiegel and G. Y. Lei, including the effects of fillers and of internal adsorption, exploring the

53

the role of segmental mobility, and establishing the general
validity of H. Frisch's theory of activated diffusion.

From the early '50's on to date, studies were conducted on
polymers with biochemical implications, e.g. the surface chemistry
of proteins and their molecular weight determination by surface
balance with E. Mishuck, followed by investigations of sickel cell
hemoglobin with J. Watson, R. Fenichel and A. Liquori, tying the
lowered solubility to changes in globin polyelectrolytic character,
and on the carcinogenicity of polymers with Dr. and Mrs. Oppenheimer,
establishing that it is the disturbance of local metabolism and not
degradation products which lead to neoplastic growth: polymers in
finely divided form were found to be practically inoccuous. There
followed synthetic, kinetic, and analytical work for over a decade
for the Surgeon General on blood plasma extenders with K. Stern,
R. Woodhams, I. A. Saad, S. Barkin, P. Frank, E. Farber and E.
Immergut, including polyvinylpyrrolidone, dextran, gelatin and related compounds, and grafting on carbohydrates and ester polymerization with K. P. Shen and S. Reegen. A logical continuation were
polymer binding and micellization studies with F. Eriksson, J.
Eliassaff, M. Schick, S. Atlas, M. Fishman and A. Scarlatti, focus-
ing on the interaction of neutral polymers and polyelectrolytes with
cosolutes, especially with surface active agents and micelle formers.
Quantitative measurements of the binding of iodine and of dyes and
detergents established a widely occurring pattern of three-state
aggregation, from molecular to micellar. A new class of polyelectrolytes arising from the adsorption of ions on neutral polymers was
uncovered. The ability of polymers to participate in micelle
formation was ascertained and related to the behavior of polymers
as reagents, dispersants, and in complex membranes. The filtration
mechanism and the permeability of modified proteninic and cell-
ulosic membranes to fluids and solutes was studied beginning 1970
with R. W. Baker, H. Strathman, and M. T. So, and the influences
of the pore structure on the flux density determined.

A long-range study sponsored first by the Air Force and later
by the National Institute for Dental Research (NIDR), begun with
H. Frisch and R. Simha in 1953, into the adsorption of polymers at
liquid-solid interfaces, has become the starting point for an ever
increasing literature. The most important results obtained in cooperation with J. Koral, R. Lauria, G. Lopatin, R. Ullman, W.
Schmidt, F. Rowland, B. Bulas, E. Rothstein, E. Chough, and J.
Wadman and also later, in conjuction with the bioadhesion studies,
with A. Kudish and H. Y. Chao, were the findings that in the absence
of strong binding forces such as between polyelectrolytes and
oppositely charged surfaces, neutral polymers from dilute solution
are invariably deposited first in conformations rather close to
those adopted by free polymer chains in solution, that is they were
sufficiently loosely held to exhibit an adsorbed layer thickness

very close to the radius of gyration of the free polymer coil; remaining responsive to any changes in the supernatant fluid, the
effects of theta-solvents, of pH, of helix-coil transitions, etc.,
could be clearly shown to occur on the surfacebound polymer. Also
established was the fact that the adsorbed polymer coils primarily
formed lightly compressed monolayers; double layers were seen only
for very stiff chains. These observations offered also new insights
into the mechanism of adhesion and of surface coatings, as expressed
in a number of summary papers.

In 1960, a series of investigations assisted by ONR was started
with R. Sabia on the viscoelastic behavior of modified elastomers.
The effects of plasticiers on PVC were shown to depend on the degree
of crystallinity of the latter and, for filled PVC, on the degree
of accumulation of the plasticizer at the interfaces. There followed the series of studies with E. Otocka on the rheology of elastomers with polar end-groups, elucidating the mechanics and kinetics
of hydrogen bonding between the chains, and further the work on
corresponding behavior of ionomers, with results which stimulated
a great deal of research elsewhere. With a different emphasis,
work with M. Boel and Z. Glaser on the aspects of rubber elasticity
related to high strains and to high speed extension including the
effects of free volume, of strain crystallization and the course
of Poisson's ratio with deformation. Volumetric studies revealed
early expansion under strain, followed by contraction as molecular
strain orientation increases. Even at extremely high rates of extension, strain induced crystallization occurs rapidly and is the
major source of the calorimetrically measured heat development.
Minor morphological differences, as between NR and NATSYN show up
by their thermomechanical responses. Work on polymer chain conformations in bulk, 1968-to date, especially on fluorinated
acrylics with B. McGarvay and W. Lee, on elastomeric dienes with
H. Mark and K. Sato, and of entanglements in GRS with R. Meyers,
followed. These experiences were reviewed, with emphasis on the
structural factors in rubber-like behavior at all phases of extension, in a chapter coauthored by T. L. Smith, on the Rupture
of Elastomers in Liebowitz's treatise "Fracture", while later
analyses focused on the effect of fillers in elastomers, the role
of interfaces in composites, failure mechanisms in plastics as a
function of morphology, and on resulting lessons for polymer engineering.

In the late '50's and continued to date, a series of studies
was undertaken for NIDR with H. Lipschitz, Ming Chan, J. Horowitz,
D. Boyer, and T. S. Chan, on bioadhesion, calcification and on
biomaterials for dental restorations. It was found that the only
satisfactory adhesion for plastic tooth restorations can be obtained
by elastomeric interliners, that for lasting bioadhesion a thin
zone of polymer-interpenetration, precipitation and/or reaction
has to be created, and that this zone should be lightly calcified.

The rates of calcification and decalcification of dentin were subjected to a detailed kinetic study, developing a new micromethod
for following the progress of de- and recalcification rates of
small bone samples. Calcification was found to be a diffusion controlled process with relatively simple kinetics, and the solubility
of the collagen-initiated mineral formation was of decisive influence readily explaining the effects of fluorine and stannous
ions; there were also indication that the nucleating site on collagen
is a phosphoprotein. A related investigation into ion binding and
complex coacervation of polyelectrolytes, especially of parent
gelatin, polystyrene sulfonate, carrageenan and acrylic acids, was
undertaken with M. Lindemann, E. A. MacMullen, C. Earl, S.
Bekelnitzky, G. Thomas, J. Katcher and T. S. Chan. Cation and anion
binding were seen to follow the Hofmeister series, polyelectrolytes
could be mutually titrated and precipitated stoichiometrically,
adsorbed ions exchanged against polyions according to a modified
mass law, while interpenetration of a polyelectrolyte into an
oppositely charged gel leads to surface precipitation as the
starting zone for further diffusion into the gel; ion binding
strongly affects the strength of dry parent gelatine and collagen
films. During the same period the solvation of the peptide group
was subjected to detailed physical chemical studies with P.
Assarsson and N. Y. Chen, finding that two water molecules are
strongly held to the =N-CO- group, with a third one bonded depending
on the peptides moiety. Polypeptides solvate on the same basis,
but in a highly individual fashion depending on amino acid residues
and sequences.

Towards the end of the '60's, work was started with the aid of
the PL 480 in cooperation with M. Paecht and the late Ahron
Katchalsky of the Weizmann Institute on the catalysis of polypeptide syntheses in aqueous solid dispersions. The most effective
catalyst was found by the Rehovot group to be montmorillonite, and
the unusual kinetics spectral molecular weight distributions, and
laws of copolymerization were established. Adsorption studies with
H. Hsu showed the nature of attachment of amino acids, their derivatives, and of polypeptides on clays to vary from ion exchange and
macromolecular coil deposition to intercalcations between stacked
clay particles. Flat ring-like, conformations of adsorbed and
growing polypeptide chains on clay surfaces, centered around in-
digenous cations which compensate for lattice defects, are most
likely responsible for the observations by the Rehovot group. In
view of the implications of this work for our understanding of
chemical evolution, this work was supported since 1973 by NASA and
is now continued with M. Halman and M. Paecht of the Weizmann Institute with respect to other potential prebiotic syntheses.

Though only about 10% of Dr. Eirich's publications deal with
rheological problems proper, e.g. the creep of PVC, the dynamic

mechanical behavior of elastomers, melt flow, abrasion high speed
extension, and the molecular mechanism of large deformations, many
other papers deal with flow as indicator of macromolecular properties and their changes. It was from concern with rheology's diffusion across almost all fields of science, that Dr. Eirich initiated
and edited from 1956 to 1969 the treatise "Rheology" to provide a
focal base as well as perspective for appreciating the importance
of rheology for other disciplines. When colloid chemistry suffered
a similar identity diffusion, he assisted E. Matijevic in co-editing
the first volumes of the series "Colloid and Surface Science".
Parallel considerations led to undertaking the coeditorship, with
A. Dietz and R. Andrews, of Plastech's symposia series on "High
Speed Testing".

Summarizing, the two common denominators of the various aspects
of Dr. Eirich's work are, on the other hand, emphasis on general
concepts underlying ongoing diverse efforts in the physics and
chemistry of macromolecules and colloids; on the other, doing
research basic to such diverse areas as adsorption of polymers and
the hydration of peptides, or the rupture of elastomers and the
calcification of bone and teeth, research which had been bypassed
by mission oriented workers but now stands to enchance the development of new research and to help the reinterpretation and inter-
gration of existing data.

Academic Degrees
Ph.D., University of Vienna, 1929
Dr. Phil. Habil., University of Vienna, 1938
M.A. Hon. Stat., University of Cambridge, 1939

Honors, Offices and Fellowships
Distinguished Professor, Polytechnic Institute, 1969; President,
Society of Rheology, 1972-73; Chairman, AAA, Gordon Research
Conferences (Polymers, 1959; Adhesion, 1961; Interfaces, 1967),
Councillor 1961-1966; Gov. Bd., AIP, '72-'74; Sigma Xi Distinguished
Service Award, 1970; Chrmn, Div. Colloid and Surface Chemistry, ACS;
Chrmn, Division of Chemistry, N.Y. Academy of Sciences; also Fellow
N.Y. Acad. Sci., Faraday Society; Visting Professor Polymer Science,
Uppsala, 1950; Visit. Unilever Prof., Dept. of Chemistry, Univ. of
Bristol, 1964-65; Dean of Research, Polytech, 1967-1970; 1974
Lecturer, Robert A. Welch Foundation; Dist. Service Award. Div.
Coll. & Surf. Chem., ACS 1975; Honorary Fellow, International
Institute Fracture Mechanics, 1976-

Professional Societies and Affiliations
American Chemical Society; American Physical Society; Society of
Rheology; Faraday Society; New York Academy of Sciences; N.Y. Zoo-
logical Society; Fellow, Geographical Society.

Publications
About 145 scientific papers and articles in professional journals;
editor of "Rheology," co-editor, "Colloid and Surface Science,"
"High Speed Testing."

Professional and Teaching Experience
Senior Research Officer, Univ. of Melbourne, 1941-43; Lecturer,
Physical Chemistry, University of Cambridge, 1943-47; Research
Associate, University of Cambridge, 1939-40; Res. Assoc., Univ. of
Vienna, 1930-1938. (See also Visiting Professorships above),
Assoc. Prof. 1947-51; Professor 1951-.

Major Fields of Interest
Polymers, Colloids, Physical Chemistry, Biomaterials, Rheology,
Science Policy.

Current Research or Other Professional Activities
Government Contracts and Committees
NASA Polypeptide Catalysis on Clays, Chemical Evolution
NIH Bioadhesives, Adsorption of Macromolecules, Rheology.
Colloid Chemistry of Calcification, Polyelectrolytes, Biopolymer
Solvation.

Selected Publications and Research Reports
Papers: "Viscoelastic Behavior of Plasticized Polyvinyl Chloride
at Large Deformation. III. The Effect of Filler," J. Polym. Sci.,
Part $\underline{A2}$, 1909-1924 (1964); co-author R. Sabia.

"On the Strength of Polymeric Materials at High Rates of Strain,"
Applied Polymer Symposia, $\underline{5}$, (1) pp. 271-300 (1965).

"Structure of Macromolecules at Liquid-Solid Interfaces," Ind. and
Eng. Chem. $\underline{57}$, 46 (1965); co-authors: F. Rowland, R. Bulas, E.
Rothstein.

"Ionic Bonding in Butadiene Copolymers. I. Network Formation," J.
Polym. Sci., Part A-2, $\underline{6}$, 921-932 (1968); co-author: E. Otocka.

"Ionic Bonding in Butadiene Copolymers. II. Stress Relaxation," J.
Polym. Sci., Part A-2, $\underline{6}$, 933-946 (1968); co-author: E. Otocka.

"The Adsorption of Solvated Macromolecules and Its Relation to the
Viscosity and Stability of Dispersions," 1st Int'l Conf. on
Hemorheology, July 1966 (Pergamon Press, 1967).

"Interaction of Water with Alkyl-Substituted Amides," Advances in
Chemistry Series, No. 84, Molecular Assoc. in Biologucal and Re-
lated Systems, American Chemical Soc., 1968; co-author: P.
Assarsson.

"Factors in Interface Conversion for Polymer Coatings," General Motors Symposium, Interface Conversion for Polymer Coatings, Ed. P. Weiss, American Elsevier, NY 1968.

"Effects of Polymer Adsorption and the Stability of Dispersions," *Amer. Chem. Soc.*, 7 (2) pp. 136-47 (1967); co-authors: J. Wadman, E. Chough.

"Thermo-Mechanics and Structure of Elastomers," PIBAL Report No. 69-38, September 1969; co-author Z. Glazer.

"NMR Studies of Fluorinated Acrylic Polymers," *J. Polym. Sci.*, Part C., No. 22, p. 1197-1206 (1969); co-authors: W. Lee, B. McGarvey.

"Properties of Amides in Aqueous Solution. I.A. VIscosity and Density Changes of Amide-Water Systems. B. An Analysis of Volume Deficiencies of Mistures Based on Molecular Size Differences (Mixing of Hard Spheres)," *J. Phys. Chem.*, $\underline{72}$, 2710 (1968) co-author: P. Assarsson.

"Isotope Effect in Gas Transport. I. Measurement of Free Volume Element in Four Polymers Above the Glass Transition Temperature," *Journal of Polymer Sci.*: Part A-2, Vol. 8, 2015-2027 (1970) co-author K. Ziegel.

"Thermal Behavior of Elastomers at High Rates of Tensile Straining," *J. Polymer Sci.*: Part C, No. 31, pp. 275-290 (1970) co-author: Z. Glazer.

"Interactions of Aqueous Poly)N-vinylpyrrolidone) with Sodium Dodecyl Sulfate. I. Equilibrium Dialysis Measurements [1a, 1b]," *J. Phys. Chem.*, 75, 3135 (1971) co-author: M.L. Fishman.

"Low Pressure Ultrafiltration of Sucrose and Raffinose Solutions with Anisotropic Membranes," *J. Phys. Chem.*, $\underline{76}$, 238 (1972) co-authors: R. W. Baker and H. Strathmann.

"On the Two-Phase Nature of the Mechanical Response of Pure and Filled Elastomers," F. R. Eirich, Proc. of the 1971 Int'l Conf. on Mechanical Behavior of Materials, Vol. III, pp. 405-418, The Soc. of Materials Science, Japan 1972.

"Molecular Mechanical Aspects of the Isothermal Rupture of Elastomers," Fracture, Vol. III, H. Liebowitz, Ed., Academic Press, 1972; co-author: T. L. Smith.

"On the Mechanism of Filler Reinforcement," International Rubber Conference, Brighton, May 1972.

"Preparation of Asymmetric Loeb-Sourirajan Membranes," Polymer
Letters Edition Vol. 11, pp. 201-205 (1973); co-authors: M. T. So,
H. Strathmann, R. W. Baker.

"NDT for and Characterization of Intrinsic Properties of Engineer-
ing Importance in Structural Materials," from "Proc. of the Inter-
disciplinary Workshop on Non-destructive Testing-Materials
Characterization, sponsored by NSF, Air Force Materials Lab, North
American Rockwell Science Center. Report No. AFML-TR-73-69, Air
Force Materials Lab, Air Force Systems Command, Wright-Patterson
Air Force Base, Ohio, April 1973.

"Calcification and Decalcification Rates in Vitro, a New Technique,"
Proceedings of Soc. for Experimental Biology and Medicine, 143,
919-924 (1973); co-author: M. S. Chan.

"Failure Modes of Elastomers," Engin. Fracture Mechanics, Vol. 5,
p. 555 (1973).

"Rubber Reinforcement in Perspective," Colloques Internatioaux du
C.N.R.S., Le Bischenberg-Obernai, No. 231, September 1973 (published 1975).

"Adhesion to Nonideal Surfaces, and to Wet Collagenous Tissue in
Particular," in "Dental Adhesive Materials" Moskowitz-Ward-
Woodbridge, Eds., Symp. Brookdale Dental Center, NYU, November
1973, p. 53.

"Polymer Engineering and Its Relevance to National Materials Development," National Science Foundation Workshop, Washington, DC,
December 1972 (1973); co-author: M. L. Williams.

"Isotope Effect in Gas Transport. II. Effect of the Glass Transition on the Size of the Free Volume Element," K. D. Ziegel, F. R.
Eirich, J. of Polymer Sci., Vol. 12, (1974) pp. 1127-1135.

"Kinetics of Demineralization and Remineralization of Dentin in
Vitro," Trans. Aidr Meeting, Atlanta, GA., 1974; co-author: D. Boyer

"Adsorption from Polymer Solutions and the Flow of Colloid-Polymer
Dispersions through Porous Media," Preprint in 48th National Col-
olid Symposium, the University of Texas at Austin, Austin, Texas,
June 24-26, 1974, p. 165 (with H. Chao, E. Chough, C. Earl, G.
Lopatin, J. Wadman).

"Thermodynamic Aspects of the Mixing of Mono-Amides and Poly-
Peptides with Water," reproduced in "Colloidal Dispersions and
Micellar Behavior" K. L. Mittal Ed., ASC Symposium Series, Vol. 9,
(1975) p. 288; co-authors: P. Assarsson and N. Y. Chen.

"Remineralization of Dentin in Vitro", D. B. Boyer and F. R. Eirich, Calcif. Tiss. Res., 21, 17-28 (1976).

"Polymers and Other Composites:, Plenary Lecture, IUPAC International Symp. Macromolecules, Jerusalem 1975, Pure & Appl. Chem., 46, 115 (1976).

"Configurational Characteristics of trans-1,4-polychloroprene: Experimental Results on the Unperturbed Dimensions and Their Temperature Coefficient:, Kyosaku Sato, F. R. Eirich and J. E. Mark, J. Polymer Sci., 14, 619-629(1976).

"Interactions of Aqueous poly)N-vinylpyrrolidone) with Sodium Dodecyl Sulfate. II. Correlation of Electric conductance and Viscosity Measurements with Equilibrium Dialysis Measurements", M. L. Fishman and F. R. Eirich, J. Phys. Chem,, 79, 2740 (1975).

"The Role of Friction and Abrasion in the Drilling of Teeth", in The Cutting Edge, ed/ S. Pearlman, DHEW Publication No. NIH 76-670 p. 1, 1976.

"Bioadhesion as an Interface Phenomenon in: "Biocompatibility of Implant Materials", D. F. Williams, Editor, Pitman (Med), London 1976.

"The Conformational Stages of Macromolecules adsorbed on Solid-Liquid Interfaces", J. Colloid and Interf. Sci., 58 (Feb) 1977.

J. D. Ferry

   Most polymer scientists who are active today will not remember
the years before rubber-like elasticity and polymer solution thermodynamics were understood, when the striking phenomena of high extensibility and strong negative deviations from thermodynamic ideality
were mysteries. My own interests in macromolecules began a few years
before those problems were solved by H. Mark, M. L. Huggins, P. J.
Flory, and their associates. I was exposed to both natural and synthetic polymers as a graduate student, and was puzzled by those and
other mysteries.

   G. S. Parks at Stanford had been studying the glass transition
in liquids of low molecular weight such as glucose, glycerol, and
3-methyl hexane; he offered me a Ph.D. problem of looking for a glass
transition in polyisobutylene (of low molecular weight), at that time
a laboratory curiosity. We found the expected transitions in heat
capacity and thermal expansion,[1] I believe the first such observation
for a synthetic polymer although it had been seen in natural rubber
the year before by Bekkedahl and Wood; and we set about to measure
the viscosity[2]. Here I encountered the phenomena of non-Newtonian
flow and viscoelasticity (creep and recovery) which have been prom-
inent in my interests ever since.

   Early years were a kind of random walk around macromolecules.
The glass transition work was preceded by a study of ultrafiltration
of proteins[3] with W. J. Elford in London and was followed by a year
with D. Spence, a rubber chemist who had been co-originator of ac-
celerators in vulcanization. We groped with cross-linking and molecular scission of natural rubber by unusual reagents before these
processes were clearly understood[4]. Thereafter, back to proteins

in the laboratory of E. J. Cohn at Harvard, I was associated with
J. L. Oncley in measurements of dielectric dispersion in protein
solutions,[5] which provided relaxation times and rotatory diffusion
coefficients.

It was these studies of the complex dielectric constant as a
function of frequency which led to the search for analogous methods
of studying viscoelasticity by measuring a complex viscosity or
elastic modulus. The first success was observation of shear wave
propagation in polymer solutions[6]. The only theoretical treatment
of wave propagation which could be found as a clue to analysis of
the measurements was in a geophysical journal[7]. From the wave
propagation the complex shear modulus and its frequency dependence
could be derived.

Polymer Systems and Viscoelastic Zones

At the University of Wisconsin since 1946, studies of viscoelasticity have evolved from concentrated polymer solutions to undiluted amorphous polymers, dilute solutions, lightly cross-linked
rubbers, glassy polymers, blends of different molecular weights,
copolymers, cross-linked rubbers with controlled network structures,
and so forth. It became evident that each type of system required
a different approach. Moreover, in amorphous polymers, the terminal, plateau, and transition zones had to be described separately.
Both dynamic (sinusoidal) and transient measurements such as creep
and stress relaxation have been utilized. The underlying theme of
this work is the relation of macromolecular dynamics--modes of
motion of polymer molecules--to mechanical and other physical properties.

Development of new experimental methods has been essential to
the progress of these studies, and has been largely due to my collaborators, especially the three physicists with whom I have been
associated: E. R. Fitzgerald,[8] M. H. Birnboim,[9] and J. L. Schrag[10].

Several specific series of studies will be reviewed briefly.

Time, Frequency, Temperature, and Concentration

In describing measurements of complex modulus (or viscosity)
at different temperatures, and in some cases at different polymer
concentrations, the variables were separated to provide the modulus
as an unspecified function of reduced frequency and a frequency
reduction factor as an unspecified function of temperature or concentration[11]. This principle, discussed earlier by H. Leaderman

and A. V. Tobolsky in connection with transient measurements, facilitates the interpretation of data and prediction of behavior
under experimentally inaccessible conditions. The form of the
temperature dependence function was found by M. L. Williams[12] to
be universal for most polymers, and with R. F. Landel we showed
how it can be related[13] to the temperature dependence of fractional free volume.

Free Volume Relations

Not only temperature dependence, but also dependence on pressure, molecular weight (in the transition zone), concentration of
diluent, and tensile strain could be described in terms of fractional free volume[14] by a generalized relation based on an empirical equation of A. K. Doolittle. This analysis is particularly
useful in locating the position of the transition zone of viscoelastic behavior on the frequency or time scale and its dependence
on those variables[15]. The most direct demonstration of the role
of free volume is found in the changes in shear relaxation times
following a temperature jump near the glass transition temperature,
as investigated with A. J. Kovacs[16].

Viscoelastic Behavior in the Transition Zone

In the range of time or frequency where the mechanical response
of an amorphous polymer varies from that of a soft rubber to that
of a hard glass, the relaxation spectrum (which determines time or
frequency dependence) has been obtained experimentally for about
two dozen polymers[17]. Its form depends somewhat on chemical structure, but these relations are still not well understood. At the
low frequency or long time end, it can be approximated by the Rouse
spectrum and a match here can furnish values of the monomeric friction coefficient, a measure of the location of the transition zone
on the time or frequency scale.

Dilute Polymer Solutions

Since molecular theories of viscoelasticity are available only
to describe the behavior of isolated polymer molecules at infinite
dilution, efforts have been made over the years for measurements
at progressively lower concentrations and it has been finally possible to extrapolate data to zero concentration. The behavior of
linear flexible macromolecules is well described by the Rouse-Zimm
theory based on a bead-spring model, except at high frequencies[18].
Effects of branching can be taken into account, at least for star-
shaped molecules. At low and intermediate frequencies, the molec-

cular topology controls the viscoelastic properties, which are independent of detailed chemical structure. At high frequencies (or
in solvents of high viscosity), just the opposite: the long-range
topology is irrelevant, but detailed chemical structure is important[19].

Rigid rod-like macromolecules follow the theory of Kirkwood-
Auer or Ullman with a single relaxation time[20]. For helical molecules with some flexibility, the relaxation spectrum includes one
relaxation time identified with rigid body rotation together with
a set of shorter times which are attributed to flexural motions[21].

Networks with Controlled Structures

A series of studies of lightly cross-linked rubbers revealed
some surprisingly slow relaxation processes which have been attributed to dangling branched structures entangled in the rubber network[22]. More recently, networks have been made with unattached
polymer species which are free to reptate through the structure
although extensively entangled[23]. Another novel type of network
is formed by cross-linking a rubbery polymer in a state of strain,
trapping the entanglements in non-random configurations. When
stress is removed, the system seeks a state of ease in which the
cross-link network and the trapped entanglement network exert
forces in opposite directions; analysis of these forces provides
the most direct evidence of the reality of entanglement networks
in uncross-linked polymers[24].

Diffusion of Small Penetrant Molecules

The diffusion of a small penetrant molecule (molecular weight
of the order of 200) in trace amounts through a rubbery polymer is
primarily determined by the mobility of the polymer molecules, and
the translational friction coefficient of the foreign molecule can
be correlated with the monomeric friction coefficient of the
polymer. In a series of studies with radioactively tagged pene-
trants, the effects of temperature, diluent, polymer structure,
thermodynamic compatibility, size of penetrant molecule, and other
variables have been investigated[25]. Results of this relatively
simple experiment can be used to make predictions of viscoelastic
behavior[26].

Polymerization of Fibrinogen

In an early series of experiments, at first unrelated to viscoelasticity studies, the mechanism of polymerization of fibrinogen

to form the three-dimensional structure of fibrin was investigated.
From studies of intermediate polymers which could be isolated by
use of suitable inhibitors, it was deduced that the polymerization
proceeds by staggered overlapping of the trinodular rod-shaped
fibrinogen units[27]. More recently, viscoelastic measurements on
the intermediate polymer in dilute solution show that it behaves
as a rigid rod. Viscoelastic properties of fibrin clots are also
being studied[28].

Current and Future Investigations

Studies of viscoelasticity of dilute solutions are being extended to polyelectrolytes and biopolymers to determine the degree of molecular flexibility of substances such as DNA and muscle
proteins. Various polymeric systems will be investigated in the
future to seek further understanding of the relations between molecular dynamics and macroscopic physical properties.

Unfortunately it has not been possible in this brief sketch to
mention all the students and associates who have made essential
contributions to work described here.

REFERENCES

1. J. D. Ferry and G. S. Parks, *J. Chem. Phys.*, 4, 70 (1936).
2. J. D. Ferry and G. S. Parks, *Physics*, 6, 356 (1935).
3. W. J. Elford and J. D. Ferry, *Biochem. J.*, 28, 650 (1934).
4. D. Spence and J. D. Ferry, *J. Amer. Chem. Soc.*, 59, 1648 (1937).
5. J. D. Ferry and J. L. Oncley, *J. Amer. Chem. Soc.*, 63, 272
   (1941).
6. J. D. Ferry, *J. Amer. Chem. Soc.*, 64, 1323 (1942).
7. D. Derjaguine, *Beitr. angew. Geophysik*, 4, 452 (1934).
8. E. R. Fitzgerald and J. D. Ferry, *J. Colloid Sci.*, 8, 1 (1953).
9. M. H. Birnboim and J. D. Ferry, *J. Applied Phys.*, 32, 2305,
   (1961).
10. J. L. Schrag and J. D. Ferry, *Faraday Symp. Chem. Soc.*, 6, 182
    (1972).
11. J. D. Ferry, *J. Amer. Chem. Soc.*, 72, 3746 (1950).
12. M. L. Williams, *J. Phys. Chem.*, 59, 95 (1955).
13. M. L. Williams, R. F. Landel, and J. D. Ferry, *J. Amer. Chem.
    Soc.*, 77, 3701 (1955).
14. J. D. Ferry and R. L. Stratton, *Kolloid-Z.*, 171, 107 (1960).
15. J. D. Ferry, in Deformation and Fracture of High Polymers, ed.
    by H. H. Kausch, J. A. Hassell, and R. I. Jaffee, Plenum Press,
    New York, 1973, p. 27.
16. A. J. Kovacs, R. L. Stratton, and J. D. Ferry, *J. Phys. Chem.*,
    67, 152 (1963).

17. J. D. Ferry, <u>Viscoelastic</u> <u>Properties</u> <u>of</u> <u>Polymers</u>, 2nd Ed.,
    Wiley, New York, 1970.
18. J. D. Ferry, *Accts. Chem. Res.*, <u>6</u>, 60 (1973).
19. J. W. M. Noordermeer, J. D. Ferry, and N. Nemoto, *Macro-
    molecules*, <u>8</u>, 672 (1975).
20. N. Nemoto, J. S. Schrag, J. D. Ferry, and R. W. Fulton,
    Biopolymers, <u>14</u>, 409 (1975).
21. T. C. Warren, J. L. Schrag, and J. D. Ferry, *Biopolymers*, <u>12</u>,
    1905 (1973).
22. N. R. Langley and J. D. Ferry, *Macromolecules*, <u>1</u>, 353 (1968).
23. O. Kramer, R. Greco, R. A. Neira, and J. D. Ferry, *J. Polymer
    Sci.*, Polymer Phys. Ed., <u>12</u>, 2361 (1974).
24. O. Kramer, R. L. Carpenter, V. Ty, and J. D. Ferry, *Macro-
    molecules*, <u>7</u>, 79 (1974).
25. S. P. Chen and J. D. Ferry, *Macromolecules*, <u>1</u>, 270 (1968).
26. C.-K. Rhee and J. D. Ferry, *J. Applied Polymer Sci.*, <u>21</u>, 467
    (1977).
27. J. D. Ferry, *Physiol. Rev.*, <u>34</u>, 753 (1954).
28. G. W. Nelb, C. Gerth, and J. D. Ferry, *Biophys. Chem.*, <u>5</u>, 377
    (1976).

P. J. Flory

I was born on 19 June, 1910, in Sterling, Illinois, of Huguenot-German parentage, mine being the sixth generation native to America. My father was Ezra Flory, a clergyman-educator; my mother, nee Martha Brumbaugh, had been a schoolteacher. Both were descended from generations of farmers in the New World. They were the first of their families of record to have attended college.

My interest in science, and in chemistry in particular, was kindled by a remarkable teacher, Carl W. Holl, Professor of Chemistry at Manchester College, a liberal arts college in Indiana, where I graduated in 1931. With his encouragement, I entered the Graduate School of the Ohio State University where my interests turned to physical chemistry. Research for my dissertation was in the field of photochemistry and spectroscopy. It was carried out under the guidance of the late Professor Herrick L. Johnston whose boundless zeal for scientific research made a lasting impression on his students.

Upon completion of my Ph.D. in 1934, I joined the Central Research Department of the DuPont Company. There it was my good fortune to be assigned to the small group headed by Dr. Wallace H. Carothers, inventor of nylon and neoprene, and a scientist of extraordinary breadth and originality. It was through the association with him that I first became interested in exploration of the fundamentals of polymerization and polymeric substances. His conviction that polymers are valid objects of scientific inquiry proved contagious. The time was propitious, for the hypothesis that polymers are in fact covalently linked macromolecules had been established by the works of Staudinger and Carothers only a few years earlier.

    A year after the untimely death of Carothers, in 1937, I
joined the Basic Science Research Laboratory of the University of
Cincinnati for a period of two years. With the outbreak of World
War II and the urgency of research and development on synthetic
rubber, supply of which was imperiled, I returned to industry,
first at the Esso (now Exxon) Laboratories of the Standard Oil
Development Company (1940-43) and later at the Research Laboratory
of the Goodyear Tire and Rubber Company (1943-48). Provision of
opportunities for continuation of basic research by these two
industrial laboratories to the limit that the severe pressures of
the times would allow, and their liberal policies on publication,
permitted continuation of the beginnings of a scientific career
which might otherwise have been stifled by the exigencies of those
difficult years.

    In the Spring of 1948 it was my privilege to hold the George
Fisher Baker Non-Resident Lectureship in Chemistry at Cornell
University. The invitation on behalf of the Department of Chemistry
had been tendered by the late Professor Peter J. W. Debye, then
Chairman of that Department. The experience of this lectureship
and the stimulating associations with the Cornell faculty led me
to accept, without hesitation, their offer of a professorship
commencing in the Autumn of 1948. There followed a most productive
and satisfying period of research and teaching. "Principles of
Polymer Chemistry," published by the Cornell University Press in
1953, was an outgrowth of the Baker Lectures.

    It was during the Baker Lectureship that I perceived a way
to treat the effect of excluded volume on the configuration of
polymer chains. I had long suspected that the effect would be
non-asymptotic with the length of the chain; that is, the perturbation of the configuration by the exclusion of one segment of the
chain from the space occupied by another would increase without
limit as the chain is lengthened. The treatment of the effect by
resort to a relatively simple "smoothed density" model confirmed
this expectation and provided an expression relating the perturbation of the configuration to the chain length and the effective
volume of a chain segment. It became apparent that the physical
properties of dilute solutions of macromolecules could not be
properly treated and comprehended without taking account of the
perturbation of the macromolecule by these intramolecular interactions. The hydrodynamic theories of dilute polymer solutions
developed a year or two earlier by Kirkwood and by Debye were
therefore reinterpreted in light of the excluded volume effect.
Agreement with a broad range of experimental information on
viscosities, diffusion coefficients and sedimentation velocities
was demonstrated soon thereafter.

    Out of these developments came the formulation of the hydro-
dynamic constant called Φ, and the recognition of the Theta point

at which excluded volume interactions are neutralized. Criteria
for experimental identification of the Theta point are easily
applied. Ideal behavior of polymers, natural and synthetic, under
Theta conditions has subsequently received abundant confirmation
in many laboratories. These findings are most gratifying. More
importantly, they provide the essential basis for rational interpretation of physical measurements on dilute polymer solutions,
and hence for the quantitative characterization of macromolecules.

In 1957, my family and I moved to Pittsburgh where I undertook to establish a broad program of basic research at the Mellon
Institute. The opportunity to achieve this objective having been
subsequently withdrawn, I accepted a professorship in the Department of Chemistry at Stanford University in 1961. In 1966, I was
appointed to the J. G. Jackson--C. J. Wood Professorship in
Chemistry at Stanford.

The change in situation upon moving to Stanford afforded the
opportunity to recast my research efforts in new directions. Two
areas have dominated the interests of my coworkers and myself since
1961. The one concerns the spatial configuration of chain molecules
and the treatment of their configuration-dependent properties by
rigorous mathematical methods; the other constitutes a new approach
to an old subject, namely, the thermodynamics of solutions. Our
investigations in the former area have proceeded from foundations
laid by Professor M. V. Volkenstein and his collaborators in the
Soviet Union, and were supplemented by major contributions of the
late Professor Kazuo Nagai in Japan. Theory and methods in their
present state of development permit realistic, quantitative
correlations of the properties of chain molecules with their
chemical constitution and structure. They have been applied to a
wide variety of macromolecules, both natural and synthetic, including polypeptides and polynucleotides in the former category. The
success of these efforts has been due in no small measure to the
outstanding students and research fellows who have collaborated
with me at Stanford during the past thirteen years. A book
entitled "Statistical Mechanics of Chain Molecules," published in
1969, summarizes the development of the theory and its applications
up to that date.

Mrs. Flory, the former Emily Catherine Tabor, and I were
married in 1936. We have three children: Susan, wife of Professor
George S. Springer of the Department of Mechanical Engineering at
the University of Michigan; Melinda, wife of Professor Donald E.
Groom of the Department of Physics at the University of Utah; and
Dr. Paul John Flory, Jr., currently a post-doctoral Research
Associate at the Medical Nobel Institute in Stockholm. We have
four grandchildren: Elizabeth Springer, Mary Springer,
Susanna Groom, and Jeremy Groom.

HONORS AND AWARDS

Joseph Sullivant Medal, The Ohio State University, 1945.
Baekeland Award, New Jersey Section, American Chemical Society,
  1947.
Sc.D. (Honorary), Manchester College (Indiana), 1950.
Colwyn Medal, Institution of Rubber Industry, Great Britain, 1954.
Nichols Medal, New York Section, American Chemical Society, 1962.
High-Polymer Physics Prize, American Physical Society, 1962.
Laurea Honoris Causa, Politecnico di Milano, 1964.
International Award in Plastics Science and Engineering, 25th
  Anniversary, Society of Plastics Engineers, 1967.
Charles Goodyear Medal, American Chemical Society, 1968.
Peter Debye Award in Physical Chemistry, American Chemical
  Society, 1969.
D.Sc., Honoris Causa, University of Manchester, England, 1969.
Sc.D. (Honorary), The Ohio State University, 1970.
Charles Frederick Chandler Medal, Columbia University, 1970.
First Award for Excellence-Chemistry, The Carborundum Company, 1971.
Cresson Medal, The Franklin Institute, 1971.
John G. Kirkwood Medal, Yale University, 1971.
J. Willard Gibbs Medal, Chicago Section, American Chemical Society,
  1973.
Priestley Medal, American Chemical Society, 1974.
Nobel Prize in Chemistry, 1974.

SPATIAL CONFIGURATION OF MACROMOLECULAR CHAINS

Nobel Lecture, December 11, 1974

The science of macromolecules has developed from primitive beginnings to a flourishing field of investigative activities within
the comparatively brief span of some forty years. A wealth of knowledge has been acquired and new points of view have illumined various
branches of the subject. These advances are the fruits of efforts
of many dedicated investigators working in laboratories spread around
the world. In a very real sense, I am before you on this occasion
as their representative.

In these circumstances, the presentation of a lecture of a
scope commensurate with the supreme honor the Royal Swedish Academy
of Sciences has bestowed in granting me the Nobel Prize for Chemistry
is an insuperable task. Rather than attempt to cover the field com-
prehensively in keeping with the generous citation by the Royal
Academy of Sciences, I have chosen to dwell on a single theme. This
theme is central to the growth of ideas and concepts concerning
macromolecules and their properties. Implemented by methods that
have emerged in recent years, researches along lines I shall attempt
to highlight in this lecture give promise of far-reaching advances
in our understanding of macromolecular substances--materials that
are invaluable to mankind.

These polymeric substances are distinguished at the molecular
level from other materials by the concatenation of atoms or groups
to form chains, often of great length. That chemical structures of
this design should occur is implicit in the multivalency manifested
by certain atoms, notably carbon, silicon, oxygen, nitrogen, sulfur,
and phosphorus, and in the capacity of these atoms to enter into
sequential combinations. The concept of a chain molecule consisting
of atoms covalently linked is as old as modern chemistry. It dates
from the origins of the graphic formula introduced by Couper in 1858
and advanced by Kekulé, Loschmidt and others shortly thereafter.
Nothing in chemical theory, either then apparent or later revealed,
sets a limit on the number of atoms that may be thus joined together.
The rules of chemical valency, even in their most primitive form,
anticipate the occurrence of macromolecular structures.

The importance of macromolecular substances, or polymers, is
matched by their ubiquity. Examples too numerous to mention abound
in biological systems. They comprise the structural materials of
both plants and animals. Macromolecules elaborated through processes of evolution perform intricate regulatory and reproductive
functions in living cells. Synthetic polymers in great variety are
familiar in articles of commerce. The prevailing structural motif
is the linear chain of serially connected atoms, groups or structural units. Departures from strict linearity may sometimes occur

through the agency of occasional branched units that impart a rami-
fied pattern to the overall structure. Linearity is predominant in
most macromolecular substances, however.

It is noteworthy that the chemical bonds in macromolecules differ in no discernible respect from those in "monomeric" compounds of
low molecular weight. The same rules of valency apply; the lengths
of bonds, e.g., C—C, C–H, C–O, etc., are the same as the corresponding bonds in monomeric molecules within limits of experimental measurement. This seemingly trivial observation has two important
implications: first, the chemistry of macromolecules is coextensive
with that of low molecular substances; second, the chemical basis
for the special properties of polymers that equip them for so many
applications and functions, both in nature and in the artifacts of
man, is not therefore to be sought in peculiarities of chemical
bonding but rather in their macromolecular constitution, specifically, in the attributes of long molecular chains.

Comprehension of the spatial relationships between the atoms
of a molecule is a universal prerequisite for bridging the connection
between the graphic formula and the properties of the substance so
constituted. Structural chemistry has provided a wealth of information on bond lengths and bond angles. By means of this information
the graphic formula, primarily a topological device, has been super-
seded by the structural formula and by the space model that affords
a quantitative representation of the molecule in three dimensions.
The stage was thus set for the consideration of rotations about
chemical bonds, i.e., for conformational organic compounds, especially cyclic ones. A proper account of bond rotations obviously
is essential for a definitive analysis of the spatial geometry of a
molecule whose structure permits such rotations.

The configuration of a linear macromolecule in space involves
circumstances of much greater complexity. A portion of such a molecule is shown schematically in Figure 1. Consecutive bonds comprising
the chain skeleton are joined at angles θ fixed within narrow limits.
Rotations φ may occur about these skeletal bonds. Each such rotation
is subject, however, to a potential determined by the character of
the bond itself and by hindrances imposed by steric interactions
between pendant atoms and groups. The number and variety of configurations (or conformations in the language of organic chemistry)
that may be generated by execution of rotations about each of the
skeletal bonds of a long chain, comprising thousands of bonds in a
typical polymer, is prodigious beyond comprehension. When the macromolecule is free of constraints, e.g., when in dilute solution, all
of these configurations are accessible. Analysis of the manner in
which such a molecule may arrange itself in space finds close analogies elsewhere in science, e.g., in the familiar problem of random
walk, in diffusion, in the mathematical treatment of systems in one
dimension, and in the behavior of real gases.

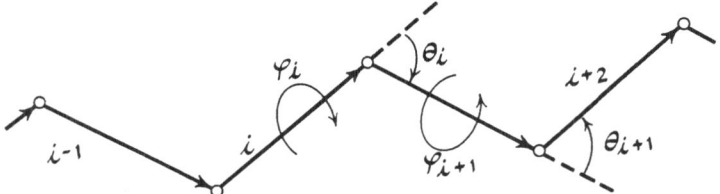

Fig. 1.   Representation of the skeletal bonds of a section of a chain molecule showing supplements θ of bond angles, and torsional rotations φ for bonds i, i + 1, etc.

Inquiry into the spatial configuration of these long-chain molecules, fascinating in itself, derives compelling motivation from its close relevancy to the properties imparted by such molecules to the materials comprising them. Indeed, most of the properties that distinguish polymers from other substances are intimately related to the spatial configurations of their molecules, these configurations being available in profusion as noted. The phenomenon of rubber-like elasticity, the hydrodynamic and thermo- dynamic properties of polymer solutions, and various optical pro- perties are but a few that reflect the spatial character of the random macromolecule. The subject is the nexus between chemical constitution and physical and chemical properties of polymeric substances, both biological and synthetic.

The importance of gaining a grasp of the spatial character of polymeric chains became evident immediately upon the establishment, ca. 1930, of the hypothesis that they are covalently linked mole- cules rather than aggregates of smaller molecules, an achievement due in large measure to the compelling evidence adduced and force- fully presented by H. Staudinger, Nobelist for 1953. In 1932 K. H. Meyer[1] adumbrated the theory of rubber-like elasticity by calling attention to the capacity of randomly coiled polymer chains to accommodate large deformations owing to the variety of con- figurations accessible to them.

W. Kuhn[2] and E. Guth and H. Mark[3] made the first attempts at mathematical description of the spatial configurations of random chains. The complexities of bond geometry and of bond rotations, poorly understood at the time, were circumvented by taking refuge in the analogy to unrestricted random flights, the theory of which had been fully developed by Lord Rayleigh. The skeletal bonds of the molecular chain were thus likened to the steps in a random walk in three dimensions, the steps being uncorrelated one to another. Restrictions imposed by bond angles and hindrances to rotation were dismissed on the grounds that they should not affect the form of the results.

For a random flight chain consisting of n bonds each of length ℓ, the mean-square of the distance r between the ends of the chain is given by the familiar relation

$$\langle r^2 \rangle = n\ell^2 \tag{1}$$

The angle brackets denote the average taken over all configurations. Kuhn[4] argued that the consequences of fixed bond angles and hindrances to rotation could be accommodated by letting several bonds of the chain molecule be represented by one longer "equivalent" bond, or step, of the random flight. This would require n to be diminished and ℓ to be increased in Eq. 1. Equivalently, one may preserve the identification of n and ℓ with the actual molecular quantities and replace Eq. (1) with

$$\langle r^2 \rangle = C\,n\ell^2, \tag{2}$$

where C is a constant for polymers of a given homologous series, i.e., for polymers differing in length but composed of identical monomeric units. The proportionality between $\langle r^2 \rangle$ and chain length expressed in Eq. (2) may be shown to hold for any random chain of finite flexibility, regardless of the structure, provided that the chain is of sufficient length and that it is unperturbed by external forces or by *effects due to excluded volume (cf. seq.)*.

The result expressed in Eq. (2) is of the utmost importance. Closely associated with it is the assertion that the density distribution W(r) of values of the end-to-end vector **r** must be Gaussian for chains of sufficient length, irrespective of their chemical structure, provided only that the structure admits of some degree of flexibility. Hence, for large n the distribution of values of r is determined by the single parameter $\langle r^2 \rangle$ that defines the Gaussian distribution.

Much of polymer theory has been propounded on the basis of the Kuhn "equivalent" random flight chain, with adjustment of n and ℓ, or of C, as required to match experimental determination of $\langle r^2 \rangle$ or of other configuration-dependent quantities. The validity of this model therefore invites critical examination. Its *intrinsic artificiality* is its foremost deficiency. Actual bond lengths, bond angles and rotational hindrances cannot be incorporated in this model. Hence, contact is broken at the outset with the features of chemical constitution that distinguish macromolecular chains of one kind from those of another. The model is therefore incapable of accounting for the vast differences in properties exhibited by the great variety of polymeric substances.

The random flight chain is patently unsuited for the treatment of constitutive properties that are configuration-dependent, e.g., dipole moments, optical polarizabilities and dichroism. Inasmuch as the contribution to one of these properties from a structural unit of the chain is a vector or tensor, it cannot be referenced to an equivalent bond that is a mere line. Moreover, the equivalent bond cannot be embedded unambiguously in the real structure.

Methods have recently been devised for treating macromolecular chains in a realistic manner. They take full account of the structural geometry of the given chain and, in excellent approximation, of the potentials affecting bond rotations as well. Before discussing these methods, however, I must direct your attention to another aspect of the subject. I refer to the notorious effect of volume exclusion in a polymer chain.

At the hazard of seeming trite, I should begin by pointing out that the chain molecule is forbidden to adopt a configuration in which two of its parts, or segments, occupy the same space. The fact is indisputable; its consequences are less obvious. It will be apparent, however, that volume exclusion vitiates the analogy between the trajectory of a particle executing a random flight and the molecular chain, a material body. The particle may cross its own path at will, but self intersections of the polymer chain are forbidden.

The effect of excluded volume must be dealt with regardless of the model chosen for representation of the chain. In practice, elimination of the effect of volume exclusion is a prerequisite to the analysis of experimental results, as I will explain in more detail later.

The closely related problems of random flights with disallowance of self intersections and of volume exclusion within long-chain molecules have attracted the attention of many theorists. A variety of mathematical techniques have been applied to the treatment of these problems, and a profusion of theories have been put forward, some with a high order of sophistication. Extensive numerical computations of random walks on lattices of various sorts also have been carried out. Convergence of results obtained by the many investigators captivated by the subject over the past quarter century seems at last to be discernible. I shall confine myself to a brief sketch of an early, comparatively simple approach to the solution of this problem.[5] The results it yields contrast with its simplicity.

Returning to the analogy of the trajectory traced by a particle undergoing a sequence of finite displacements, we consider only those trajectories that are free of intersections as being acceptable for the chain molecule. Directions of successive steps may or may not be correlated, i.e., restrictions on bond

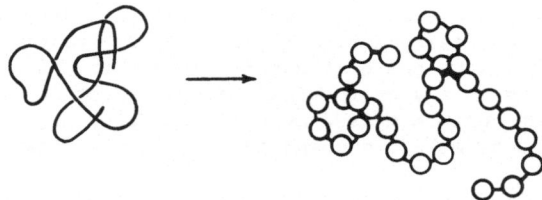

Fig. 2. The effect of excluded volume. The configuration on the
left represents the random coil in absence of volume exclusion,
the chain being equivalent to a line in space. In the sketch on
the right, the units of the chain occupy finite domains from which
other units are excluded, with the result that the average size of
the configuration is increased.

angles and rotational hindrances may or may not be operative; this
is immaterial with respect to the matter immediately at hand.
Obviously, the set of eligible configurations will occupy a larger
domain, on the average, than those having one or more self intersections. Hence, volume exclusion must cause $<r^2>$ to increase.
The associated expansion of the spatial configuration is illustrated
in Fig. 2. Other configuration-dependent quantities may be affected
as well.

This much is readily evident. Assessment of the magnitude of
the perturbation of the configuration and its dependence on chain
length require a more penetrating examination.

The problem has two interrelated parts: (i) the mutual
exclusion of the space occupied by segments comprising the chain
tends to disperse them over a larger volume, and (ii) the con-
comitant alteration of the chain configuration opposes expansion
of the chain. Volume exclusion (i) is commonplace. It is pre-
valent in conventional dilute solutions and in real gases, molecules of which mutually exclude one another. In the polymer chain
the same rules of exclusion apply, but treatment of the problem is
complicated by its association with (ii).

Pursuing the analogies to dilute solutions and gases, we
adopt a "smoothed density" or "mean field" model. The segments of
the chain, x in number, are considered to pervade a volume V, the
connections between them being ignored insofar as part (i) is
concerned. The segment need not be defined explicitly; it may be
identified with a repeating unit or some other approximately isometric portion of the chain. In any case, x will be proportional
to the number n of bonds; in general $x \neq n$, however. For simplicity, we may consider the segment density ρ to be uniform throughout the volume V; that is, $\rho = x/V$ within V and $\rho = 0$ outside of V.

This volume should be proportional to $\langle r^2 \rangle^{3/2}$, where $\langle r^2 \rangle$ is the mean-square separation of the ends of the chain averaged over those configurations *not disallowed by excluded volume interactions*. Accordingly, we let

$$V = A\langle r^2 \rangle^{3/2}, \tag{3}$$

where A is a numerical factor expected to be of the order of magnitude of unity.

It is necessary to digress at this point for the purpose of drawing a distinction between $\langle r^2 \rangle$ for the chain perturbed by the effects of excluded volume and $\langle r^2 \rangle_0$ for the unperturbed chain in the absence of such effects. If α denotes the factor by which a linear dimension of the configuration is altered, then

$$\langle r^2 \rangle = \alpha^2 \langle r^2 \rangle_0 \tag{4}$$

Equation (2), having been derived without regard for excluded volume interactions, should be replaced by

$$\langle r^2 \rangle_0 = C \, n\ell^2, \tag{2'}$$

where C reaches a constant value with increase in n for any series of finitely flexible chains.

The smoothed density within the domain of a linear macromolecule having a molecular weight of 100,000 or greater (i.e., $n > 1000$) is low, only on the order of one percent or less of the space being occupied by chain segments. For a random dispersion of the segments over the volume V, encounters in which segments overlap are rare in the sense that few of them are thus involved. However, the expectation that such a dispersion is entirely free of overlaps between any pair of segments is very small for a long chain. The attrition of configurations due to excluded volume is therefore severe.

In light of the low segment density, it suffices to consider only binary encounters. Hence, if β is the volume excluded by a segment, the probability that an arbitrary distribution of their centers within the volume V is free of conflicts between any pair of segments is

$$P_{(i)} \approx \prod_{i=1}^{x} (1 - i\beta/V) \approx \exp(-\beta x^2/2V). \tag{5}$$

Introduction of Eq. (3) and (4) gives

$$P_{(i)} = \exp(-\beta x^2/2A\langle r^2 \rangle_0^{3/2}\alpha^3) \tag{6}$$

or, in terms of the conventional parameter z defined by

$$z = (3/2\pi)^{3/2} \, (<r^2>_0/x)^{-3/2} x^{1/2} \beta, \tag{7}$$

$$P_{(i)} = \exp[-2^{1/2}(\pi/3)^{3/2} A^{-1} z \alpha^{-3}]. \tag{8}$$

Since $<r^2>_0$ is proportional to x for long chains (see Eq. (2')), z depends on the square-root of the chain length for a given series of polymer homologs.

We require also the possibility $P_{(ii)}$ of a set of configurations having the average density corresponding to the dilation α^3 relative to the probability of a set of configurations for which the density of segments corresponds to $\alpha^3 = 1$. For the former, the mean-squared separation of the ends of the chain is $<r^2>$; for the latter it is $<r^2>_0$. The distribution of chain vectors r for the unperturbed chain is approximately Gaussian as noted above. That is to say, the probability that r falls in the range r to r+dr is

$$W(r)dr = \text{Const} \exp(-3r^2/2<r^2>_0)dr, \tag{9}$$

where dr denotes the element of volume. The required factor is the ratio of the probabilities for the dilated and the undilated sets of configurations. These probabilities, obtained by taking the products of W(r)dr over the respective sets of configurations, are expressed by W(r) according to Eq. (9) with r^2 therein replaced by the respective mean values, $<r^2>$ and $<r^2>_0$, for the perturbed and unperturbed sets. Bearing in mind that the volume element dr is dilated as well, we thus obtain

$$P_{(ii)} = [(dr)/(dr)_0] \exp[-3(<r^2>-<r^2>_0)/2<r^2>_0]$$
$$= \alpha^3 \exp[-(3/2) \, (\alpha^2-1)]. \tag{10}$$

The combined probability of the state defined by the dilation α^3 is

$$P_{(i)} P_{(ii)} = \alpha^3 \exp[-2^{1/2}(\pi/3)^{3/2} A^{-1} z \alpha^{-3} - (3/2) \, (\alpha^2-1)]. \tag{11}$$

Solution for the value of α that maximizes this expression gives

$$\alpha^5 - \alpha^3 = 2^{1/2}(\pi/3)^{3/2} A^{-1} z. \tag{12}$$

Recalling that z is proportional to $x^{1/2}\beta$ according to Eq. (7), one may express this result alternatively as follows:

$$\alpha^5 - \alpha^3 = B x^{1/2} \beta, \tag{12'}$$

where $B = (<r^2>_0/x)^{-3/2}(2A)^{-1}$ is a constant for a given series of polymer homologs.

In the full treatment[5,6] of the problem along the lines
sketched briefly above, the continuous variation of the mean segment density with distance from the center of the molecule is
taken into account, and the appropriate sums are executed over all
configurations of the chain. The squared radius of gyration s^2,
i.e., the mean-square of the distances of the segments from their
center of gravity, is preferable to r^2 as a parameter with which
to characterize the spatial distribution.[7] Treatments carried out
with these refinements affirm the essential validity of the result
expressed by Eq. (12) or (12'). They show conclusively[7,8] that
the form of the result should hold in the limit of large values of
$\beta x^{1/2}$, i.e., for large excluded volume and/or high chain length,
and hence for $\alpha \gg 1$. In this limit, $(\alpha^5-\alpha^3)/z=1.67$ according to
H. Fujita and T. Norisuye.[8] For $\alpha < \sim 1.4$, however, this ratio
decreases, reaching a value of 1.276 at $\alpha = 1$.[8,9]

The general utility of the foregoing result derived from the
most elementary considerations is thus substantiated by elaboration
and refinement of the analysis, the quantitative inaccuracy of Eqs.
(12) and (12') in the range $1.0 < \alpha \leqslant 1.4$ notwithstanding. The
relationship between α and the parameter z prescribed by these
equations, especially as refined by Fujita and Norisuye,[8] appears
to be well supported by experiment.[10,11]

The principal conclusions to be drawn from the foregoing
results are the following: the expansion of the configuration
due to volume exclusion increases with chain length *without limit*
for $\beta > 0$; for very large values of $\beta x^{1/2}$ relative to $(<r^2>_0/x)^{3/2}$
it should increase as the 1/10 power of the chain length. The
sustained increase of the perturbation with chain length reflects
the fact that interactions between segments that are remote in
sequence along the chain are dominant in affecting the dimensions
of the chain. It is on this account that the excluded volume
effect is often referred to as a long-range interaction.[9-12]

The problem has been treated by a variety of other procedures.[9-12] Notable amongst these treatments is the self-
consistent field theory of S. F. Edwards.[12] The asymptotic
dependence of α on the one-tenth power of the chain length, and
hence the dependence of $<r^2>$ on $n^{6/5}$ for large values of the
parameter z, has been confirmed.[12]

The dilute solution is the milieu chosen for most physico-
chemical experiments conducted for the purpose of characterizing
polymers. The effect of excluded volume is reflected in the properties of the polymer molecule thus determined, and must be taken into
account if the measurements are to be properly interpreted. The
viscosity of a dilute polymer solution is illustrative. Its useful-

ness for the characterization of polymers gained recognition largely
through the work of Staudinger and his collaborators.

Results are usually expressed as the intrinsic viscosity $[\eta]$
defined as the ratio of the increase in the relative viscosity
η_{rel} by the polymeric solute to its concentration c in the limit
of infinite dilution. That is,

$$[\eta] = \lim_{c \to 0} [\eta_{rel}-1)/c]$$

the concentration c being expressed in weight per unit volume.
The increment in viscosity due to a polymer molecule is proportional to its hydrodynamic volume, which in turn should be
proportional to $<r^2>^{3/2}$ for a typical polymer chain. Hence, $\eta_{rel}-1$
should be proportional to the product of $<r^2>^{3/2}$ and the number
density of solute molecules given by c/M where M is the molecular
weight. It follows that

$$[\eta] = \Phi<r^2>^{3/2}/M, \tag{13}$$

where Φ is a constant of proportionality.[6,13] Substitution from
Eq. (4) and rearrangement of the result gives

$$[\eta] = \Phi(<r^2>_0/M^{3/2}M^{1/2}\alpha^3 \tag{13'}$$

The ratio $<r^2>_0/M$ should be constant for a series of homologs
of varying molecular weight, provided of course that the molecular
weight, and hence the chain length, is sufficiently large.

If the excluded volume effect could be ignored, the intrinsic
viscosity should vary proportionally to $M^{1/2}$. Since, however, α
increases with M, a stronger dependence on M generally is observed.
Often the dependence of $[\eta]$ on molecular weight can be represented
in satisfactory approximation by the empirical relation

$$[\eta] = KM^a \tag{14}$$

where $0.5 \leqslant a \leqslant 0.8$. Typical results are shown by the upper sets
of data in Figs. 3 and 4 for polystyrene dissolved in benzene[14]
and for poly(methyl methacrylate) in methyl ethyl ketone,[15]
respectively. Values of α^3 are in the range 1.4 to 5. At the
asymptote for chains of great length and large excluded
volume β, the exponent a should reach 0.80 according to the treatment given above. Although this limit is seldom reached within
the accessible range of molecular weights, the effects of excluded
volume can be substantial. They must be taken into account in
the interpretation of hydrodynamic measurements.[13,16] Otherwise,
the dependences of the intrinsic viscosity and the translational

friction coefficient on molecular chain length are quite incomprehensible.

Measurement of light scattering as a function of angle, a method introduced by the late P. Debye, affords a convenient means for determining the mean-square radius of gyration. Small-angle scattering of x-rays (and lately of neutrons) offers an alternative for securing the same information. From the radius of gyration one may obtain the parameter $<r^2>$ upon which attention is focused here. The results are affected, of course, by the perturbation due to excluded volume. Inasmuch as the perturbation is dependent on the solvent and temperature, the results directly obtained by these methods are not intrinsically characteristic of the macromolecule. Values obtained for $<r^2>$ from the intrinsic viscosity by use of Eq. (13), or by other methods, must also be construed to be jointly dependent on the macromolecule and its environment.

If the factor α were known, the necessary correction could be introduced readily to obtain the more substantive quantities, such as $<r^2>_0$ and $<s^2>_0$ that characterize the macromolecule itself and are generally quite independent of the solvent. Evaluation of α according to Eq. (11) and (12) would require the excluded volume β. This parameter depends on the solvent in a manner that eludes prediction. Fairly extensive experimental measurements are required for its estimation, or for otherwise making correction for the expansion α.

All these difficulties are circumvented if measurements on the polymer solution are conducted under conditions such that the effects of excluded volume are suppressed. The resistance of atoms to superposition cannot, of course, be set aside. But the consequences thereof can be neutralized. We have only to recall that the effects of excluded volume in a gas comprising real molecules of finite size are exactly compensated by intermolecular attractions at the Boyle temperature (up to moderately high gas densities). At this temperature the real gas masquerades as an ideal one.

For the macromolecule in solution, realization of the analogous condition requires a relatively poor solvent in which the polymer segments prefer self-contacts over contacts with the solvent. The incidence of self-contacts may then be adjusted by manipulating the temperature and/or the solvent composition until the required balance is established. Carrying the analogy to a real gas a step further, we require the excluded volume integral for the interaction between a pair of segments to vanish; that is, we require that $\beta=0$. This is the necessary and sufficient condition.[5,6,13]

As already noted, estimation of the value of β is difficult; the prediction of conditions under which β shall precisely vanish would be even more precarious. However, the "Theta point," so-called, at which this condition is met is readily identified with high accuracy by any of several experimental procedures. An excluded volume of zero connotes a second virial coefficient of zero, and hence conformance of the osmotic pressure to the celebrated law of J. H. van't Hoff. The Theta point may be located directly from osmotic pressure determinations, from light scattering measured as a function of concentration, or from deter-mination of the precipitation point as a function of molecular weight.[6,13]

The efficacy of this procedure, validated a number of years ago with the collaboration of T. G. Fox, W. R. Krigbaum, and others,[13,17,18] is illustrated in Figs. 3 and 4 by the lower plots of data representing intrinsic viscosities measured under ideal, or Theta conditions.[6] The slopes of the lines drawn through the lower sets of points are exactly 1/2, as required by Eq. (13') when $\beta = 0$ and hence $\alpha = 1$. The excellent agreement here illustrated has been abundantly confirmed for linear macromolecules of the widest variety, ranging from polyisobutylene and polyethylene to polyribonucleotides.[19] At the Theta point the mean-square chain vector $\langle r^2 \rangle_0$ and the mean-square radius of gyration $\langle s^2 \rangle_0$ invariably are found to be proportional to chain length.

A highly effective strategy for characterization of macro-molecules emerges from these findings. By conducting experiments at the Theta point, the disconcerting (albeit interesting!) effects of excluded volume on experimentally measured quantities may be eliminated. Parameters (e.g., $\langle r^2 \rangle_0$ and $\langle s^2 \rangle_0$) are thus obtained that are characteristic of the molecular chain. They are found to be virtually independent of the nature of the "Theta solvent" selected. Having eliminated the effects of long range inter-actions, one may turn attention to the role of short range features: structural geometry, bond rotation potentials, and steric interactions between near-neighboring groups. It is here that the influences of chemical architecture are laid bare. If the marked differences in properties that distinguish the great variety of polymeric substances, both natural and synthetic, are to be rationally understood in fundamental, molecular terms, this must be the focus of future research.

Rigorous theoretical methods have recently become available for dealing realistically with short-range features peculiar to a given structure. Most of the remainder of this lecture is devoted to a brief overview of these methods. Although the field is com-paratively new and its exploration has only begun, space will not permit a digest of the results already obtained.

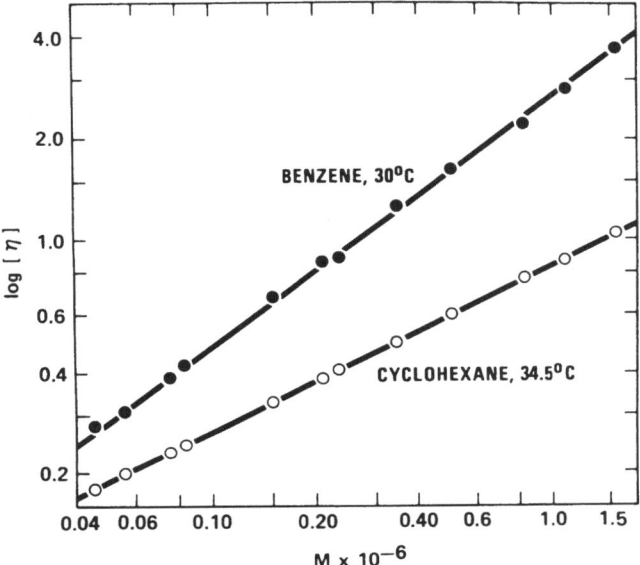

Fig. 3. Intrinsic viscosities of polystyrene fractions plotted against their molecular weights on logarithmic scales in accordance with Eq. (14). The upper set of data was determined in benzene, a good solvent for this polymer. The lower set of data was determined in cyclohexane at the Theta point. The slopes of the lines are a = 0.75 and 0.50, respectively. From the results of Altares, Wyman and Allen.[14]

The broad objective of the methods to which we now turn attention is to treat the structure and conformations accessible to the chain molecule in such a manner as will enable one to calculate configuration-dependent quantities and to average them over all conformations, or spatial configurations, of the unperturbed chain. The properties under consideration are constitutive; they represent sums of contributions from the individual units, or chemical groupings, comprising the chain. In addition to $\langle r^2 \rangle_0$ and $\langle s^2 \rangle_0$, they include: mean-square dipole moments; the optical anisotropies underlying strain birefringence, depolarized light scattering and electric birefringence; dichroism; and the higher moments, both scalar and tensor, of the chain vector $\mathbf{r}$. Classical statistical mechanics provides the basis for evaluating the configurational averages of these quantities. Since bond lengths and bond angles ordinarily may be regarded as fixed, the bond rotations φ are the variables over which averaging must be carried out. The procedure rests on the *rotational isomeric state scheme*, the

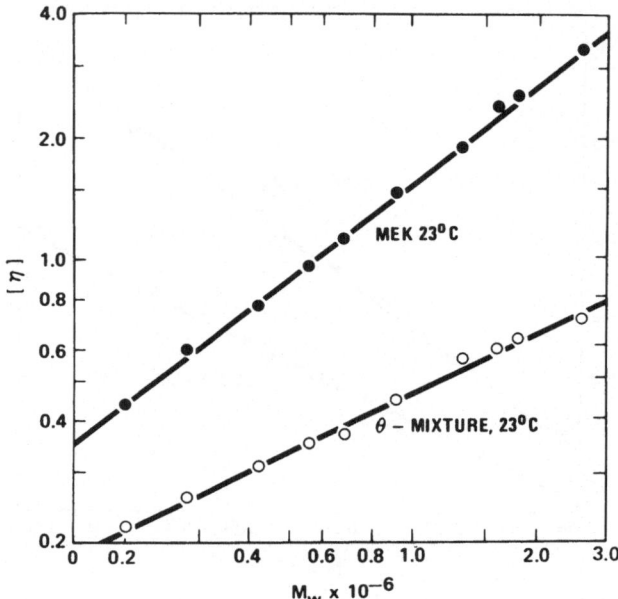

Fig. 4. Intrinsic viscosities of fractions of poly(methyl methacrylate) according to Chinai and Samuels[15] plotted as in Fig. 3. The
upper set of points was measured in methyl ethyl ketone, a good
solvent. The lower set was determined in a mixture of methyl
ethyl ketone and isopropanol at the Theta point. Slopes are a =
0.79 and 0.50, respectively.

foundations for which were set forth in large measure by M. V.
Volkenstein[20] and his colleagues[21] in Leningrad in the late
1950's and early 1960's. It is best explained by examples.

Consider rotation about an internal bond of an n-alkane chain.
As is now well established,[22,23] the three staggered conformations,
trans(t), gauche-plus(g^+) and its mirror image, gauche-minus(g^-),
are of lower energy than the eclipsed forms. The t and g^- conformations of n-butane are shown in Fig. 5. The energies of the
eclipsed conformations separating t from g^+ and t from g^- are
about 3.5 kcal. mol^{-1} above the energy of the trans conformation.
Hence, in good approximation, it is justified to consider each bond
to occur in one of three *rotational isomeric states* centered near
(but not necessarily precisely at) the energy minima associated
with the three staggered conformations.[20-24] The gauche minima lie
at an energy of about 500 cal. mol^{-1} above trans. Each of the
former is therefore disfavored compared to the latter by a

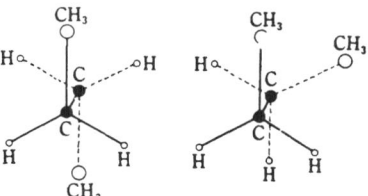

Fig. 5. Two of the staggered conformations for n-butane: trans
on the left and gauche-minus on the right.

"statistical weight" factor we choose to call $\sigma \approx \exp(-E_g/RT)$, where
E_g is about 500 cal. mol^{-1}; thus, $\sigma \approx 0.5$ at T = 400 K.

A complication arises from the fact that the potentials affecting bond rotations usually are neighbor dependent, i.e., the potential affecting ϕ_i depends on the rotations ϕ_{i-1} and ϕ_{i+1}. Bond
rotations cannot, therefore, be treated independently.[20,21,24,25]
The source of this interdependence in the case of an n-alkane chain
is illustrated in Fig. 6 showing a pair of consecutive bonds in
three of their nine conformations. In the conformations tt, tg^+,
g^+t, tg^-, and g^-t, the two methylene groups pendant to this pair
of bonds are well separated. For gauche rotations g^+g^+ and g^-g^-
of the same hand (Fig. 6b), these groups are proximate but not
appreciably overlapped. Semi-empirical calculations[21,24,26,27]
show the intramolecular energy for these two equivalent conformations to be very nearly equal to the sum (ca. 1000 cal. mol^{-1}) for
two well-separated gauche bonds; i.e., the interdependence of the
pair of rotations is negligible. In the remaining conformations,
g^+g^- and g^-g^+, the steric overlap is severe (Fig. 6c). It may be
alleviated somewhat by compromising rotations, but the excess
energy associated therewith is nevertheless about 2.0 kcal. mol^{-1}.
Hence, a statistical weight factor $\omega \approx \exp(-2000/RT)$ is required
for each such pair.[24,26,28] Inspection of models in detail shows
that interactions dependent upon rotations about three, four of
five consecutive bonds are disallowed by interferences of shorter
range and hence may be ignored.[24] It suffices therefore to consider first neighbors only.

The occurrence of interactions that depend on pairs of skeletal
bonds is the rule in chain molecules. In some of them, notably in
vinyl polymers, such interactions may affect most of the conformations. Hence, interdependence of rotations usually plays a major
role in determining the spatial configuration of the chain. The
rotational isometric state approximation, whereby the continuous
variation of each ϕ is replaced by discrete states, provides the
key to mathematical solution of the problem posed by rotational
interdependence.[20,21,24,25]

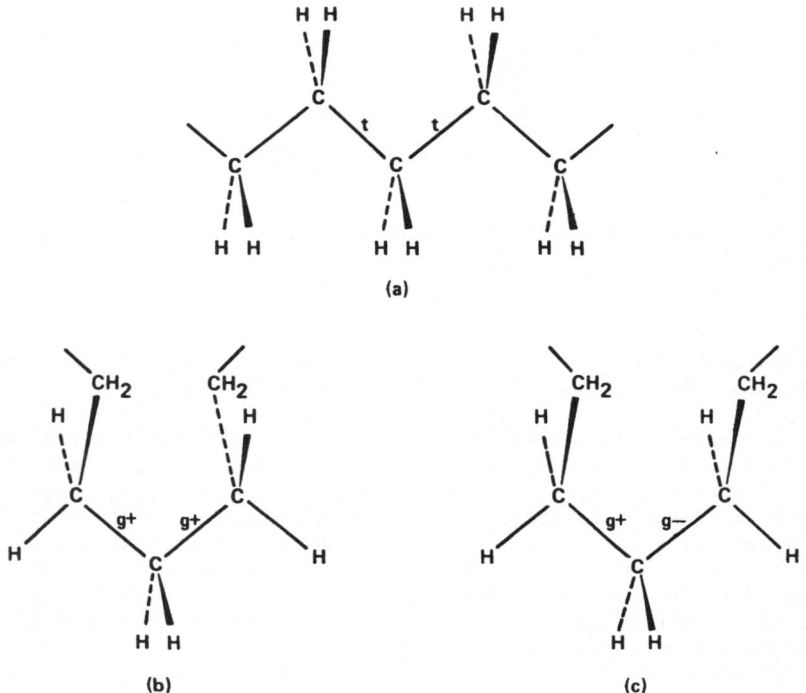

Fig. 6. Conformations for a pair of consecutive bonds in an n-alkane chain: (a), tt; (b), g^+g^+; (c), g^+g^-. Wedged bonds project forward from the plane of the central bonds, dashed bonds project behind this plane.

It is necessary therefore to consider the bonds pairwise consecutively, and to formulate a set of statistical weights for bond i that take account of the state of bond i-1. These statistical weights are conveniently presented in the form of an array, or matrix, as follows:

$$
U_i = \begin{bmatrix} u^{tt} & u^{tg^+} & u^{tg^-} \\ u^{g^+t} & u^{g^+g^+} & u^{g^+g^-} \\ u^{g^-t} & u^{g^-g^+} & u^{g^-g^-} \end{bmatrix}_i , \tag{15}
$$

where the rows are indexed in the order t, g^+, g^- to the state of bond i-1, and the columns are indexed to the state of bond i in the same order. According to the analysis of the alkane chain conformations presented briefly above, U_i takes the form,[24,26,28]

$$U_i = \begin{bmatrix} 1 & \sigma & \sigma \\ 1 & \sigma & \sigma\omega \\ 1 & \sigma\omega & \sigma \end{bmatrix}_i \qquad (16)$$

for any bond $1 < i < n$.

A conformation of the chain is specified in the rotational isometric state approximation by stipulation of the states for all internal bonds 2 to n-1 inclusive; e.g., by $g^+ttg^-g^-$, etc. Owing to the three-fold symmetry of the terminal methyl groups of the alkane chain, rotations about the terminal bonds are inconsequential and hence are ignored. The statistical weight for the specified conformation of the chain is obtained by selecting the appropriate factor for each bond from the array (15) according to the state of this bond and of its predecessor, and taking the product of such factors for all bonds 2 to n-1. In the example above this product is $u^{g^+}\,u^{g^+t}\,u^{tt}\,u^{tg^-}\,u^{g^-g^-}$, etc. It will be obvious that the first superscripted index in one of the factors u must repeat the second index of its predecessor since these indices refer to the same bond.

The configuration partition function, representing the sum of all such factors, one for each conformation of the chain as represented by the scheme of rotational isomeric states, is

$$Z = \sum_{\text{all states}} u_2 u_3 \ldots u_1 \ldots u_{n-1}, \qquad (17)$$

where the subscripts are serial indexes. Each u_i must be assigned as specified above. The sum, which extends over all ordered combinations of rotational states, may be generated identically as the product of the arrays U_i treated as matrices. That is, according to the rules of matrix multiplication

$$Z = \prod_{i=1}^{n} U_i, \qquad (18)$$

where $U_1 =$ row $(1, 0, 0)$ and $U_n =$ column $(1, 1, 1)$. Matrix multiplication generates products precisely of the character to which attention is directed at the close of the preceding paragraph. Serial multiplication of the statistical weight matrices generates this product for each and every conformation of the chain, and Eq. (18) with the operators U_1 and U_n appended gives their sum.

The foregoing procedure for evaluation of Z is a minor variant of the method of H. A. Kramers and G. H. Wannier[29] for

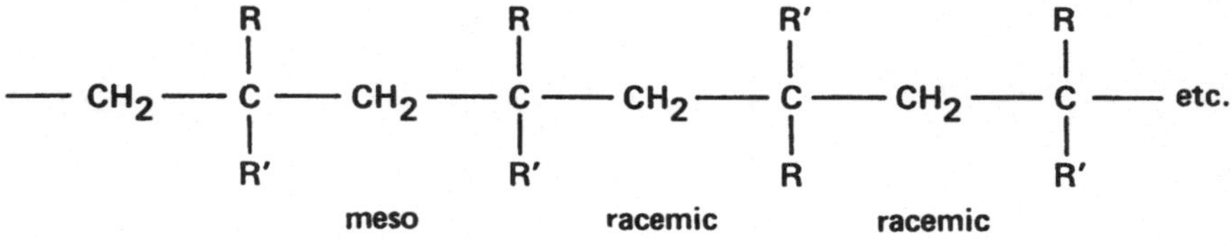

$$
\begin{array}{ccccccc}
& R & & R & & R' & & R \\
& | & & | & & | & & | \\
-CH_2- & C & -CH_2- & C & -CH_2- & C & -CH_2- & C & - \text{ etc.} \\
& | & & | & & | & & | \\
& R' & & R' & & R & & R' \\
& \text{meso} & & \text{racemic} & & \text{racemic} &
\end{array}
$$

Fig. 7. A vinyl polymer chain shown in projection in its planar (fully extended) conformation. If the substituents R and R' differ (e.g., if $R = C_6H_5$ and $R' = H$ as in polystyrene), diastereomeric dyads must be distinguished as indicated for the stereo-chemical structure shown.

treating a hypothetical one-dimensional ferromagnet or lattice. A number of interesting characteristics of the chain molecule can be deduced from the partition function by application of familiar techniques of statistical mechanics. I shall resist the temptation to elaborate these beyond mentioning two properties of the molecule that may be derived directly from the partition function, namely, the incidences of the various rotational states and combinations thereof, and the equilibrium constants between isomeric structures of the chain in the presence of catalysts effectuating their inter-conversion. Vinyl polymers having the structure depicted in Fig. 7 with $R' \neq R$ afford examples wherein the study of equilibria between various diastereomeric forms arising from the local chirality of individual skeletal bonds has been especially fruitful.[30]

Consider the evaluation of a configuration-dependent property for a given configuration, or conformation, of the chain. Since the configuration is seldom "given", the problem as stated is artificial. Its solution, however, is a necessary precursor to the ultimate goal, which is to obtain the average of the property over all configurations. A property or characteristic of the chain that will serve for illustration is the end-to-end vector r. Suppose we wish to express this vector with reference to the first two bonds of the chain. For definiteness, let a Cartesian coordinate system be affixed to these two bonds with its X_1-axis along the first bond and its Y_1-axis in the plane of bonds 1 and 2, as shown in Fig. 8. The vector r is just the sum $\sum_{i=1}^{n} l_i$ of all of the bond vectors l_i, each expressed in this reference frame.

In order to facilitate the task of transforming every bond vector to the reference frame affiliated with the first bond, it is helpful to define a reference frame for each skeletal bond of

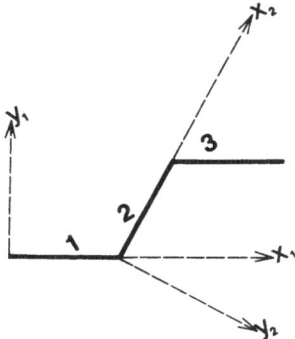

Fig. 8. Specification of the coordinate axes affixed to each of the first two bonds of the chain: X_1Y_1 for bond 1 and X_2Y_2 for bond 2.

the chain. For example, one may place the axis X_i along bond i, the Y_i-axis in the plane of bonds i-1 and i, and choose the Z_i-axis to complete a right-handed Cartesian system. Let T_i symbolize the transformation that, by premultiplication, converts the representation of a vector in reference frame i+1 to its representation in the preceding reference frame i. Then bond i referred to the initial reference frame is given by

$$T_1 T_2 \ldots T_{i-1} l_i,$$

where l_i is presented in reference frame i. The required sum is just

$$r = \sum_{i=1}^{n} T_1 \ldots T_{i-1} l_i. \tag{19}$$

This sum of products can be generated according to a simple algorithm. We first define a "generator" matrix A_i as follows[31,32]

$$A_i = \begin{bmatrix} T_i & l_i \\ & \\ 0 & 1 \end{bmatrix}, \qquad 1 < i < n, \tag{20}$$

together with the two terminal matrices

$$A_1 = [T_1 \ l_1], \tag{21}$$

$$A_n = \begin{bmatrix} 1_n \\ 1 \end{bmatrix}. \qquad (22)$$

In these equations T_i is the matrix representation of the transformation specified above and 0 is the null matrix of order 1 X 3. The desired vector r is generated identically by taking the serial product of the A's; i.e.,

$$r = \prod_{i=1}^{n} A_i, \qquad (23)$$

as may easily be verified from the elementary rules of matrix multiplication. Each generator matrix A_i depends on the length of bond i and, through T_i, on both the angle θ_i between bonds i and i+1 and on the angle of rotation ϕ_i about bond i (see Fig. 1).

In order to obtain the average of r over all configurations of the chain, it is necessary to evaluate the sum over all products of the kind given in Eq. (23) with each of them multiplied by the appropriate statistical weight for the specified configuration of the chain; see Eq. (17). That is,

$$\langle r \rangle_0 = Z^{-1} \Sigma u_2 u_3 \ldots u_{n-1} A_1 A_2 \ldots A_n, \qquad (24)$$

where the sum includes all configurations. This sum can be generated by serial multiplication of matrices defined as follows:

$$a_i = \begin{bmatrix} u_{tt}A_t & u_{tg^+}A_{g^+} & u_{tg^-}A_{g^-} \\ u_{g^+t}A_t & u_{g^+g^+}A_{g^+} & u_{g^+g^-}A_{g^-} \\ u_{g^-t}A_t & u_{g^-g^+}A_{g^+} & u_{g^-g^-}A_{g^-} \end{bmatrix}_i, \quad 1 < i < n, \qquad (25)$$

$$a_1 = [A_1 \quad 0 \quad 0], \qquad (26)$$

$$a_n = \text{column} \ (A_n, A_n, A_n). \qquad (27)$$

Then[31]

$$\langle r \rangle_0 = Z^{-1} \prod_{i=1}^{n} a_i. \qquad (28)$$

The matrix a_i comprises the elements of U_i (see Eq. (15)) joined with the A matrix for the rotational state of bond i as prescribed by the column index. It will be apparent that serial multiplication of the a_i according to Eq. (28) generates the statistical weight factor $u_2 u_3 \ldots u_{n-1}$ for every configuration of the chain in the same way that these factors are generated by serial multiplication of the statistical weight matrices U_i in Eq. (18). Simultaneously, Eq. (28) generates the product of A's (see Eq. (23)) that produces the vector r for each configuration thus weighted. The resulting products of statistical weights and of A's are precisely the terms required by Eq. (24). The terminal factors in Eq. (28) yield their sum.

With greater mathematical concision[31,32]

$$a_i = (U_i \otimes E_3) \| A_i \|, \qquad 1 < i < n, \tag{29}$$

$$a_1 = U_1 \otimes A_1, \tag{30}$$

$$a_n = U_n \otimes A_n, \tag{31}$$

where E_3 is the identity matrix of order three, $\otimes$ signifies the direct product, and $\| A_i \|$ denotes the diagonal array of the matrices $A_i{}^t$, $A_i g^+$, and $A_i g^-$.

A characteristic of the chain commanding greater interest is the quantity $\langle r^2 \rangle_0$ introduced in earlier discussion. For a given configuration of the chain, r^2 is just the scalar product of r with itself, i.e.,

$$r^2 = r \cdot r = \sum_{i=1}^{n} \ell^2_i + 2 \sum\sum_{i<j} 1_i \cdot 1_j \tag{32}$$

If each bond vector 1_i is expressed in its own reference frame i, then

$$r^2 = \sum_{1}^{n} \ell_i{}^2 + 2 \sum\sum_{i<j} 1_i{}^T T_i T_{i+1} \ldots T_j{-1} 1_j, \tag{33}$$

where $1_i{}^T$ is the transposed, or row form of vector 1_i. These sums can be evaluated by serial multiplication of the generator matrices[24,33]

$$G_i = \begin{bmatrix} 1 & 2\ell^T T & \ell^2 \\ 0 & T & 1 \\ 0 & 0 & 1 \end{bmatrix}_i, \qquad 1 < i < n. \tag{34}$$

That is,

$$r^2 = \prod_{1}^{n} G_i \tag{35}$$

where G_1 has the form of the first row, and G_n that of final

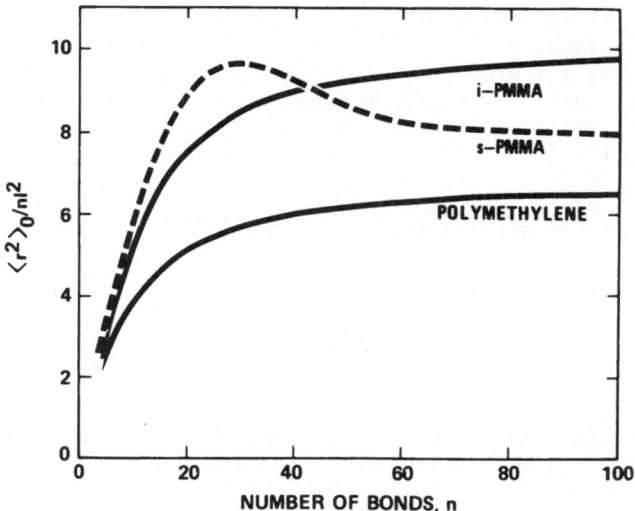

Fig. 9.   Characteristic ratios $\langle r^2\rangle_0/n\ell^2$ plotted against the number of bonds n in the chain for polymethylene, and for isotactic and syndiotactic poly(methyl methacrylate)'s.   From the calculations of Abe, Jernigan and Flory[26] and of Yoon.[34]

column of Eq. (34).   Evaluation of $\langle r^2\rangle_0$ proceeds exactly as set forth above for $\langle r\rangle_0$.[32,33]

The foregoing method enjoys great versatility.   The chain may be of any specified length and structure.   If it comprises a variety of skeletal bonds and repeat units, the factors entering into the serial products have merely to be fashioned to introduce the characteristics of the bond represented by each of the successive factors.   The mathematical methods are exact; the procedure is free of approximations beyond that involved in adoption of the rotational isometric state scheme.   With judicious choice of rotational states, the error here involved is generally within the limits of accuracy of basic information on bond rotations, nonbonded interactions, etc.

Other molecular properties that may be computed by straight-forward adaptation of these methods[24,32] include the higher scalar moments $\langle r^4\rangle_0$, $\langle r^6\rangle_0$, etc.; the moment tensors formed from r; the radius of gyration $\langle s^2\rangle_0 = (n+1)^{-2} \sum_{ij} \langle r^2_{ij}\rangle$; the optical polariz-ability anistropy as manifested in depolarized light scattering, in strain birefringence and in electric birefringence;   x-ray scattering at small angles;   and NMR chemical shifts.

For illustration, characteristic ratios $\langle r^2 \rangle_0 / n\ell^2$ are plotted in Fig. 9 against the numbers n of bonds for n-alkanes and for isotactic and syndiotactic poly(methyl methacrylate), or PMMA. Isotactic PMMA is represented by the formula in Fig. 7 with R = $COOCH_3$ and R' = CH_3 and with all dyads of the meso form, i.e., with R occurring consistently above (or below) the axis of the chain. In the syndiotactic stereoisomer, the substituents R and R' alternate from one side to the other, all dyads being racemic.

For the alkane and the isotactic PMMA chains the characteristic ratios increase monotonically with chain length, approaching asymptotic values for n ≈ 100 bonds. This behavior is typical. For syndiotactic PMMA, however, the characteristic ratio passes through a maximum at intermediate values of n, according to these computations by D. Y. Yoon.[34] This behavior can be traced[34] to the inequality of the skeletal bond angles in PMMA in conjunction with the preference for tt conformations in the syndiotactic chain.[35] The maximum exhibited in Fig. 9 for this polymer is thus a direct consequence of its constitution. This peculiarity manifests itself in the small angle scattering of x-rays and neutrons by predominantly syndiotactic PMMA of high molecular weight.[36] Scattering intensities are enhanced at angles corresponding, roughly, to distances approximating $\langle r^2 \rangle_0^{1/2}$ at the maximum in Fig. 9. This enhancement, heretofore considered anomalous, is in fact a direct consequence of the structure and configuration of syndiotactic PMMA.

It is thus apparent that subtle features of the chemical architecture of polymeric chains are manifested in their molecular properties. Treatment in terms of the artificial models much in use at present may therefore be quite misleading.

The analysis of the spatial configurations of macromolecular chains presented above is addressed primarily to an isolated molecule as it exists, for example, in a dilute solution. On theoretical grounds, the results obtained should be equally applicable to the molecules as they occur in an amorphous polymer, even in total absence of a diluent. This assertion follows unambiguously from the statistical thermodynamics of mixing of polymer chains,[5,6,37] including their mixtures with low molecular diluents. It has evoked much skepticism, however, and opinions to the contrary have been widespread. These opposing views stem primarily from qualitative arguments to the effect that difficulties inherent in the packing of long chains of consecutively connected segments to space-filling density can only be resolved either by alignment of the chains in bundle arrays, or by segregation of individual molecules in the form of compact globules. In either circumstance, the chain configuration would be altered drastically.

Whereas dense packing of polymer chains may appear to be a distressing task, a thorough examination of the problem leads to the firm conclusion that macromolecular chains whose structures offer sufficient flexibility are capable of meeting the challenge without departure or deviation from their intrinsic proclivities. In brief, the number of configurations the chains may assume is sufficiently great to guarantee numerous combinations of arrangements in which the condition of mutual exclusion of space is met throughout the system as a whole. Moreover, the task of packing chain molecules is not made easier by partial ordering of the chains or by segregating them.[6,37] Any state of organization short of complete abandonment of disorder in favor of creation of a crystalline phase offers no advantage, in a statistical-thermodynamic sense.

Theoretical arguments aside, experimental evidence is compelling in showing the chains to occur in random configurations in amorphous polymers, and further that these configurations correspond quantitatively with those of the unperturbed state discussed above.[38] The evidence comes from a variety of sources: from investigations on rubber elasticity, chemical cyclization equilibria, thermodynamics of solutions, and, most recently, from neutron scattering studies on protonated polymers in deuterated hosts (or vice versa).[39] The investigations last mentioned go further. They confirm the prediction made twenty-five years ago that the excluded volume perturbation should be annulled in the bulk amorphous state.[5] The excluded volume effect is therefore an aberration of the dilute solution, which, unfortunately, is the medium preferred for physicochemical characterization of macromolecules.

Knowledge gained through investigations, theoretical and experimental, on the spatial configuration and associated properties of random macromolecular chains acquires added significance and importance from its direct, quantitative applicability to the amorphous state. In a somewhat less quantitative sense, this knowledge applies to the intercrystalline regions of semicrystalline polymers as well. It is the special properties of polymeric materials in amorphous phases that render them uniquely suited to many of the functions they perform both in biological systems and in technological applications. These properties are intimately related to the nature of the spatial configurations of the constituent molecules.

Investigation of the conformations and spatial configurations of macromolecular chains is motivated therefore by considerations that go much beyond its appeal as a stimulating intellectual exercise. Acquisition of a thorough understanding of the subject must be regarded as indispensable to the comprehension of rational connections between chemical constitution and those properties that

render polymers essential to living organisms and to the needs
of man.

REFERENCES

1. Meyer, K. H., von Susich, G., and Valkó, E., *Kolloid-Z*, 59, 208 (1932).
2. Kuhn, W., *Kolloid-Z*, 68, 2 (1934).
3. Guth, E., and Mark, H., *Monatsch.*, 65, 93 (1934).
4. Kuhn, W., *Kolloid-Z*, 76, 258 (1936); 87, 3 (1939).
5. Flory, P. J., *J. Chem. Phys.*, 17, 303 (1949).
6. Flory, P. J., Principles of Polymer Chemistry, Cornell University Press, Ithaca, NY, 1953.
7. Flory, P. J., and Fisk, S., *J. Chem. Phys.*, 44, 2243 (1966).
8. Fujita, H., and Norisuye, T., *J. Chem. Phys.*, 52, 115 (1971).
9. Fixman, M., *J. Chem. Phys.*, 23, 1656 (1955).
10. Yamakawa, H., Modern Theory of Polymer Solutions, Harper and Row, New York, 1971.
11. Yamakawa, H., *Pure and Appl. Chem.*, 31, 179 (1972).
12. Edwards, S. F., *Proc. Phys. Soc.*, (London), 85, 613 (1965).
13. Fox, T. G., Jr., and Flory, P. J., *Phys. and Coll. Chem.*, 53, 197 (1949). Flory, P. J., and Fox, T. G., Jr., *J. Polymer Sci.*, 5, 745 (1950); *J. Amer. Chem. Soc.*, 73, 1904 (1951).
14. Altares, T., Wyman, D. P., and Allen, V. R., *J. Polymer Sci.*, A, 2, 4533 (1964).
15. Chinai, S. N., and Samuels, R. J., *J. Polymer Sci.*, 19, 463 (1956).
16. Mandelkern, L., and Flory, P. J., *J. Chem. Phys.*, 20, 212 (1952); Mandelkern, L., Krigbaum, W. R., and Flory, P. J., ibid., 20, 1392 (1952).
17. Fox, T. G., Jr., and Flory, P. J., *J. Amer. Chem. Soc.*, 73, 1909, 1915 (1951).
18. Krigbaum, W. R., Mandelkern, L., and Flory, P. J., *J. Polymer Sci.*, 9, 381 (1952). Krigbaum, W. R., and Flory, P. J., ibid, 11, 37 (1953).
19. Eisenberg, H., and Felsenfeld, G., *J. Mol. Biol.*, 30, 17 (1967). Inners, L. D., and Felsenfeld, G., ibid., 50, 373 (1970).
20. Volkenstein, M. V., Configurational Statistics of Polymeric Chains, translated from the Russian ed., 1959, by S. N. and M. J. Timasheff, Interscience, New York, 1963.
21. Birshtein, T. M. and Ptitsyn, O. B., Conformations of Macromolecules, translated from the Russian ed., 1964, by S. N and M. J. Timasheff, Interscience, New York, 1966.
22. Pitzer, K. S., *Discussions Faraday Soc.*, 10, 66 (1951).
23. Mizushima, S., *Structure of Molecules and Internal Rotation*, Academic Press, New York, 1954.
24. Flory, P. J., Statistical Mechanics of Chain Molecules, Interscience Publishers, New York, 1969.

25. Gotlib, Yu. Ya., *Zh. Fiz Tekhn*, <u>29</u>, 523 (1959). Birshtein,
     T. M., and Ptitsyn, O. B., ibid., <u>29</u>, 1048 (1959). Lifson,
     S., *J. Chem. Phys.*, <u>30</u>, 964 (1959). Nagai, K., ibid., <u>31</u>,
     1169 (1959), Hoeve, C. A., ibid., <u>32</u>, 888 (1960).
26. Abe, A., Jernigan, R. L., and Flory, P. J., *J. Amer. Chem.
     Soc.*, <u>88</u>, 631 (1966).
27. Scott, R. A., and Scheraga, H. A., *J. Chem. Phys.*, <u>44</u>, 3054
     (1966).
28. Hoeve, C. A. J., *J. Chem. Phys.*, <u>35</u>, 1266 (1961).
29. Kramers, H. A., and Wannier, G. H., *Phys. Rev.*, <u>60</u>, 252 (1941).
30. Williams, A. D., and Flory, P. J., *J. Amer. Chem. Soc.*, <u>91</u>,
     3111, 3118 (1969). Flory, P. J., and Pickles, C. J., *J. Chem.
     Soc.*, Faraday Trans. II, <u>69</u>, 632 (1973). Suter, U. W.,
     Pucci, S., and Pino, P., *J. Amer. Chem. Soc.*, <u>97</u>, 1018 (1975).
31. Flory, P. J., *Proc. Nat. Acad. Sci.*, <u>70</u>, 1819 (1973).
32. Flory, P. J., *Macromolecules*, <u>7</u>, 381 (1974).
33. Flory, P. J., and Abe, Y., *J. Chem. Phys.*, <u>54</u>, 1351 (1971).
34. Yoon, D. Y., unpublished results, Laboratory of Macromolecular
     Chemistry, Stanford University.
35. Sundararajan, P. R., and Flory, P. J., *J. Amer. Chem. Soc.*,
     <u>96</u>, 5025 (1974).
36. Kirste, R. G., and Kratky, O., *Z. Physik, Chem. Neue Folge*,
     <u>31</u>, 363 (1962). Kirste, R. G., *Makromol. Chem.*, <u>101</u>, 91 (1967).
     Kirste, R. G., Kruse, W. A., and Ibel, K., *Polymer*, <u>16</u>, 120
     (1975).
37. Flory, P. J., *Proc. Royal Soc.*, A, <u>234</u>, 60 (1956). Flory,
     P. J., *J. Polym. Sci.*, <u>49</u>, 105 (1961).
38. Flory, P. J., *Pure & Applied Chem., Macromolecular Chem.*, <u>8</u>,
     1-15 (1972).
39. Kirste, R. G., Kruse, W. A., and Schelten, J., *Makromol. Chem.*,
     <u>162</u>, 299 (1972). Benoit, H., Decker, D., Higgins, J. S.,
     Picot, C., Cotton, J. P., Farnoux, B., Jannink, G., and Ober,
     R., *Nature, Physical Sciences*, <u>245</u>, 13 (1973). Ballard, D. G.
     H., Wignall, G. D., and Schelten, J., *Eur. Polymer J.*, <u>9</u>, 965
     (1973); ibid, <u>10</u>, 861 (1974). Fischer, E. W., Leiser, G.,
     and Ibel, K., *Polymer Letters*, <u>13</u>, 39 (1975).

M. L. Huggins

Macromolecules in Crystals and Glasses

In 1920 I began applying G.N. Lewis' theory of atomic and molecular structures, with some modifications and extensions of my own[1,2], to the structures of crystals[7]. I soon found that many crystals of known atomic arrangement were 3-dimensional macromolecules, with the atomics in each crystal all bonded together by ordinary chemical (electronpair) bonds. With the aid of published x-ray data I deduced the previously unknown structure of quartz (SiO_2), showing it to be of this type[9]. I discovered that many inorganic crystals (and graphite) were composed of sheet molecules[6,13,21,25]. An x-ray study of mercuric iodide crystals[23] showed them to be of this sort. According to the theory, crystals of S, Se or Te should be composed of either ring molecules or chain macromolecules. I showed[21,25,29] that "metallic" Se and Te agreed with the latter. Each molecule is a linear polymer with a spiral (helical) structure.

Applying my concept of the hydrogen bond[1,2] to the structure of ice, I correctly predicted[21,25,205] that it would be found to be a 3-dimensional "hydrogen-bond polymer".

The atomic arrangements in thousands of other crystalline substances have since been determined. With few exceptions, these structures conform to the simple ideas and assumptions that I used in the early 1920's. As predicted from those concepts, many silicate crystals, for example, contain macroanions: 1-, 2-, or 3-dimensional, depending on the chemical composition[186,187,197].

The same theoretical ideas, plus some generalizations deduced from known crystal structures, lead to useful conclusions about the structures and properties of glasses[64-69,79,81,91,93-95,97,99,107,112-116].

Conformations in Polymers

In 1922[205,214], I showed that the organic chemists' principle of "free rotation" about single bonds should be modified. I concluded that, in a normal alkane chain for example, an extended zigzag chain, being energetically more stable than other conformations, would exist in crystals, but that other conformations, having only slightly higher energies, would also be present in the gases and liquids (including solutions). This concept explained the alternation in melting points and some other properties of n-alkanes and their simple derivatives.

In my crystal structure studies I kept trying to deduce and apply general structural principles[7,29,83,160,163,178]. One of these principles was that like atoms or atomic groups tend to be surrounded by close neighbors in like manner. Another was that the units (molecules and atoms) tend to be as close-packed as possible, without having strong repulsions between them. (In general, strong repulsions occur whenever interatomic distances are shorter than the sums of the "non-bonded radii" or "van der Waals radii"[35,182].) These principles led me to conclude that polymeric chain molecules in crystals should either have an extended zigzag structure or a spiral (helical) structure. Using this idea, with experimental fiber axis repeat distances and approximately known bond angles and interatomic distances (conforming to my tables of approximately additive atomic radii[8,17,40]), I deduced the helical molecular structures for several polymers, including fibrous sulfur and polyoxymethylene[105]. Many later structure analyses (on stereoregular polymers, polypeptides, etc.) have corroborated my theoretical conclusions.

The Structures of Biopolymers

In cellulose, glucose rings are connected together through oxygen bridges to form a zigzag structure[25]. I have postulated[57] that the zigzag arrangement is stabilized by hydrogen bonds connecting one hydroxyl oxygen of each ring with an ether oxygen in the next ring. The other hydroxyls also form hydrogen bonds, chiefly to hydroxyls in other chains (and to water molecules, if any are present in the regions between the chains).

In starch, the orientations of the bonds bridging between adjacent glucose rings are such as to make a zigzag structure, like that in cellulose, impossible. Unbranched starch (amylose) molecules therefore (in accordance with my postulates) form helical structures. I have shown[155,159] that the x-ray data for one of the helical forms agree with a helix of a certain size, with consecutive turns connected through hydrogen bonds in a reasonable way.

The early work of Astbury on keratin, the hair protein, led him to conclude that in β-keratin (stretched hair) the polypeptide chains are nearly fully extended, but in the unstretched form (α-keratin) they are "folded". Applying structural principles, I deduced reasonable structures, for both forms, that were in agreement with most of the meager x-ray data than available[83,155,161]. In both forms I assumed NHO hydrogen bonds -- between different chains in the β-form and consecutive turns of helical chains in the α-form. It was then impossible to determine which of several reasonable structures (for the molecules in either form) is correct. No one else has since reported any structure, for either form, that satisfies all the experimental requirements.

I have recently returned to the keratin structure problem using established structural principles, with x-ray and other data from the literature, I have deduced a structure pattern for α-keratin that appears satisfactory from all standpoints[226,227]. Helical polypeptide chains, with intrachain NHO hydrogen bonds, are grouped into helically staggered "3-stacks", in which each chain is shifted in the axial direction about 5.1 A. relative to its neighbors. (This shift accounts nicely for the previously unexplained strong 5.1 A. meridional x-ray reflection.) The magnitude of the shift is presumably that needed to give optimum contact interactions between the chains[150,163,178,183].

The 3-stacks are packed together, like cylinders, to give approximate close-packing of the 3-stacks and also of the individual polypeptide helices[150,226,227]. There is a slow twisting of the individual chains and of the 3-stacks, reversing in direction when certain stability limits are reached. There are disulfide cross-links, probably producing an alternation of zones with many cross-links with zones of few crosslinks. These structural features should be studied by polymer scientists interested in making better textile fibers[227].

Again considering β-keratin, I have deduced a new structure[227], that appears to be better than any previously proposed. It needs further testing, but I believe it to be correct. Staggered 3-stacks remain, but the individual polypeptide chains are nearly fully extended. Intrachain hydrogen bonds in α-keratin have been replaced by interchain hydrogen bonds between different chains in the same 3-stack. The 3-stacks are reoriented in such a way as to produce hexagonal channels, into which all the R groups of the peptide units are directed.

With these new structures, the $\alpha \rightarrow \beta$ transition is much simpler and more reasonable than with any of the previously proposed structures.

Stereoregularity

In 1943 I learned that the polymerization of styrene at different temperatures (T. Alfrey, A. Bartovics and H. Mark, J. Am. Chem. Soc. 65, 2319 - 1943) had produced products with different physical properties. I explained this result[101] as due to differences in the degree of regularity of steric orientation of the phenyl groups attached to the asymmetric carbon atoms. I was unable to follow up this idea experimentally, but Ziegler, Natta and others have shown that it was correct.

Later, with the aid of unpublished experimental research by Dr. A. L. Geddes, I developed a theory of the mechanism of Ziegler-Natta catalysis[162,164]. Later still, as a preliminary to planned experimental research related to my theory, Sakurada, Anderson and I made an NMR study[171] of solutions of a series of simple compounds of Al, Ti, CH_3, and Cl.

Condensation Polymers

In 1922 I read a paper (R. Adams, J.E. Bullock and W.C. Wilson, J. Amer. Chem. Soc. 44, 521 - 1922) reporting that several condensation reactions between simple bifunctional molecules (e.g., benzidene and terephthaldehyde) gave small ring products. No molecular weights were determined, since the products were all insoluble in ordinary organic solvents and melted, with decomposition, at unexpectedly high temperatures. I believed them to be polymeric "string molecules"[214], but the authors had apparently not considered that possibility. Checking the literature on other similar condensations, I found many products that (to me) were obviously polymeric, but were reported as small rings.

I did no experimental work in this field, but when I presented my interpretation of these results in 1930 at a meeting of the American Chemical Society, I learned about the researches of Carothers and Staudinger.

Coordination Polymers

Starting experimental research on polymer synthesis in 1936, I made a few elastomeric coordination polymers. This research was stopped, because of the difficulty of obtaining experimental assistance[214].

Much later, I initiated the synthesis (S. Kanda and Y. Saito, Bull. Chem. Soc. Japan 30, 192 - 1957) of some rodlike coordination

polymers, in which dihydroxyquinonate groups alternate with bivalent
metal atoms having coordination numbers of four. Still later, with
Eding and other colleagues at Stanford Research Institute, I made
other products of this class and also, using trivalent 6-coordinating
metals, some amorphous, highly porous, semiconducting network
polymers[173,179,195,207,225]. I predict that further extension of
this research will lead to very interesting results. With appropriate techniques, it should also be possible to make crystalline 2-
dimensional sheet polymers, with holes of uniform, predetermined
size.

I had planned research on various other types of coordination
polymers[255], using reagents and procedures expected to avoid the
poor results that most other researchers in this field have obtained, but I was never successful in obtaining the necessary intra-
company approval or external financial assistance for the work.

Nomenclature

A 1951 Report on Nomenclature in the Field of Macromolecules[125]
and a 1961 Report on Nomenclature Dealing with Steric Regularity
in High Polymers[165], both prepared by subcommissions of the IUPAC
Commission on Macromolecules under my chairmanship, have, in my
opinion, been useful. They have served as bases[192] for more recent reports and recommendations by the present Commission on
Macromolecular Nomenclature.

Entropy-Dependent Properties

In the 1930's several scientists presented evidence showing
that rubber elasticity is essentially an entropy phenomenon,
related to the change in randomness of location of the rubber segments when the material is extended. On this basis they derived
relationships between the initial elastic modulus, the average
chain length between network junctions, etc. The basic idea
appealed to me. It was a natural extension of my 1922 theory of
conformations in simple molecules.

I was interested in the dependence on chain length (and so on
molecular weight) of various properties of linear polymers and
other chain molecules. Many of these properties depend on both
energies and entropies. I deduced some theoretically reasonable
general relationships and tested them with data on normal alkanes
[59,123]. Then I calculated theoretically the entropy of long chain
molecules in the gaseous state, assuming random kinking[61].

To relate elastic properties to molecular compositions and
energy changes associated with conformational changes, I derived
a theory of rubberlike elasticity for a simple hypothetical model,

showing how the stress-strain behavior depends on the changes in energy when the conformations in the small rearrangeable structural units change[111,118].

Solution Viscosities

I admired Staudinger's fine researches in the polymer field, but I could not accept his claim that the "Staudinger Law" -- proportionality between intrinsic viscosity and molecular weight -- could be explained on the assumption of rodlike chain molecules.

Extending a theoretical treatment by Werner Kuhn, I deduced an equation for the intrinsic viscosity of solutions of chain molecules, assuming perfectly random kinking and "free draining" (i.e., that the flow of solvent past each segment is unaffected by the presence of other segments in the vicinity). I obtained Staudinger's relation[54,56,57]. Departures from this relation at high molecular weights result primarily from departures from the free-draining approximation.

For the initial concentration dependence of the viscosity, my theory[78] gave a simple equation involving a constant, often referred to as "the Huggins constant".

Thermodynamic Properties of Solutions

It became evident in the 1930's that the inconsistent and often absurd molecular weights obtained by applying the previously accepted equations for the concentration dependence of osmotic pressure, freezing point depression, etc., resulted from the inapplicability of these equations to mixtures of molecules of very different size. A new theoretical equation for the entropy of mixing was obviously needed.

I derived such an equation and tested it on various solutions for which appropriate experimental data were in the literature[73-76,82,87,89,117,161]. Flory (J. Chem. Phys. $\underline{9}$, 660 - 1941; $\underline{10}$, 51 - 1942) independently derived a similar equation, equivalent to mine at the infinite dilution limit.

Our theoretical equations were strictly applicable only at concentrations approaching infinite dilution, but they have often been used, without good theoretical or experimental justification, for nondilute solutions. For this purpose, a function[90] (now called chi, χ), that is necessarily constant as infinite dilution is approached, is assumed to have the same constant value at other concentrations.

Since 1941, I have made repeated attempts[119,138,161] to derive better theoretical equations for nondilute solutions. During the

past few years I have spent much of my time on this problem. I
have derived new relationships that seem to be fundamentally sound,
give good agreement with accurate experimental measurements, and
involve parameters that measure pertinent molecular and intermolecular properties[190,191,193,194,198,202-204,206,208,210,211,
216-220].

My new theory is applicable to mixtures of small molecules, as
well as to polymer solutions. Using extremely precise experimental
data from the literature, I find excellent agreement, checking the
basic assumptions of my theory[222-224]. With these checks, I am
now refining my theory for polymer solutions, dealing especially
with the temperature and concentration dependence, relations of
the parameters to molecular compositions, and the prediction of
parameters from data on the pure components and other related
systems. A new treatment of the thermodynamic properties of solutions of oligomers is being prepared for publication.

Forecasts

In a number of lectures and published papers during recent
years, I have discussed certain areas of polymer science (and sometimes other fields) in which I predict interesting and important
future developments[195,199,207,221,225,227]. Although I cannot
expect to be much involved in these developments, I hope that these
suggestions, based on my own background of knowledge and experience,
will guide others into explorations that will be interesting and
lead to a better understanding of the structure and properties of
matter.

PUBLICATIONS OF M. L. HUGGINS

1. *Sci.*, 55, 459 (1922).
2. *J. Phys. Chem.*, 26, 601 (1922).
3. *J. Phys. Chem.*, 26, 833 (1922).
4. *J. Amer. Chem. Soc.*, 44, 1607 (1922).
5. *Sci.*, 55, 679 (1922).
6. *J. Amer. Chem. Soc.*, 45, 264 (1923).
7. *J. Amer. Chem. Soc.*, 44, 1841 (1922).
8. *Phys. Rev.*, 19, 346 (1922).
9. *Phys. Rev.*, 19, 363 (1922).
10. *Phys. Rev.*, 19, 354 (1922).
11. *Phys. Rev.*, 19, 369 (1922).
12. *Phys. Rev.*, 21, 509 (1923).
13. *Am. J. Sci.*, 5, 303 (1923).
14. S. B. Hendricks and M. L. Huggins, *J. Amer. Chem. Soc.*, 48,
    164, (1926).
15. J. Field, II, and M. L. Huggins, *Sci.*, 63, 433 (1926).
16. *Phys. Rev.*, 27, 286 (1926).

17. *Phys. Rev.*, <u>28</u>, 1086 (1926).
18. *J. Chem. Education*, <u>3</u>, 1110 (1926).
19. *J. Chem. Education*, <u>3</u>, 1254 (1926).
20. *J. Chem. Education*, <u>3</u>, 1426 (1926).
21. *J. Chem. Education*, <u>4</u>, 73 (1927).
22. *J. Chem. Education*, <u>4</u>, 220 (1927).
23. P. I. Magill and M. L. Huggins, *J. Amer. Chem. Soc.*, <u>49</u>, 2357 (1927).
24. *J. Opt. Soc. Amer.*, <u>14</u>, 55 (1927).
25. *Scientific Monthly*, <u>32</u>, 140 (1931).
26. *Phys. Rev.*, <u>37</u>, 447 (1931).
27. *J. Am. Chem. Soc.*,
28. *J. Phys. Chem.*, <u>35</u>, 1216 (1931).
29. *J. Phys. Chem.*, <u>35</u>, 1270 (1931).
30. *Phys. Rev.*, <u>37</u>, 1177 (1931).
31. *J. Amer. Chem. Soc.*, <u>53</u>, 3190 (1931).
32. *J. Amer. Chem. Soc.*, <u>55</u>, 3823 (1931).
33. B. A. Noble and M. L. Huggins, *Am. Min.*, <u>16</u>, 519 (1931).
34. G. O. Frank and M. L. Huggins, *Am. Min.*, <u>16</u>, 580 (1931).
35. *Chem. Rev.*, <u>10</u>, 427 (1932).
36. *Rev. Sci. Inst.*, <u>4</u>, 10 (1933).
37. *Am. Min.*, <u>18</u>, 455 (1933).
38. *Z. Krist.*, <u>(A)86</u>, 384 (1933).
39. J. E. Mayer and M. L. Huggins, *J. Chem. Phys.*, <u>1</u>, 643 (1933).
40. Linus Pauling and M. L. Huggins, *Z. Krist.*, <u>(A)87</u>, 205 (1934).
41. *Electrical Engineering*, <u>53</u>, 851 (1934).
42. *J. Chem. Phys.*, <u>3</u>, 473 (1935).
43. *J. Am. Chem. Soc.*, <u>58</u>, 694 (1936).
44. *J. Chem. Educ.*, <u>13</u>, 160 (1936).
45. *J. Chem. Phys.*, <u>4</u>, 308 (1936).
46. *J. Phys. Chem.*, <u>40</u>, 723 (1936).
47. *J. Org. Chem.*, <u>1</u>, 407 (1936).
48. *J. Chem. Phys.*, <u>5</u>, 143 (1937); <u>15</u>, 212 (1937).
49. *J. Chem. Phys.*, <u>5</u>, 201 (1937).
50. *J. Phys. Chem.*, <u>41</u>, 299 (1937).
51. *Nature*, <u>139</u>, 550 (1937).
52. *Zeit. Krist.*, <u>(A)96</u>, 384 (1937).
53. *J. Chem. Phys.*, <u>5</u>, 527 (1937).
54. *J. Phys. Chem.*, <u>42</u>, 911 (1938).
55. *J. Am. Chem. Soc.*, <u>61</u>, 755 (1939).
56. *J. Phys. Chem.*, <u>43</u>, 439 (1939).
57. *J. App. Physics*, <u>10</u>, 700 (1939).
58. *J. Am. Chem. Soc.*, <u>61</u>, 2982 (1939).
59. *J. Phys. Chem.*, <u>43</u>, 1083 (1939).
60. *Synthetic Organic Chemicals*, <u>12</u>, No. 2 (1939).
61. *J. Chem. Physics*, <u>8</u>, 181 (1940).
62. *Introduction to the Symposium*, *Chem. Reviews*, <u>26</u>, 141 (1940).
63. *J. Chem. Physics*, <u>8</u>, 598 (1940).
64. *J. Chem. Physics*, <u>8</u>, 641 (1940).
65. *J. Amer. Chem. Soc.*, <u>62</u>, 2248 (1940).

66. *J. Opt. Soc. Amer.*, <u>30</u>, 420 (1940).
67. *J. Opt. Soc. Amer.*, <u>30</u>, 495 (1940).
68. *J. Opt. Soc. Amer.*, <u>30</u>, 514 (1940).
69. *Ind. Eng. Chem.*, <u>32</u>, 1433 (1940).
70. *J. Am. Chem. Soc.*, <u>63</u>, 66 (1941).
71. *J. Am. Chem. Soc.*, <u>63</u>, 116 (1941).
72. *J. Am. Chem. Soc.*, <u>63</u>, 916 (1941).
73. *J. Chem. Physics*, <u>9</u>, 440 (1941).
74. *Ann. N.Y. Acad. Sciences*, <u>41</u>, 1 (1942).
75. *J. Phys. Chem.*, <u>46</u>, 151 (1942).
76. *J. Am. Chem. Soc.*, <u>64</u>, 1712 (1942).
77. *Ann. Rev. of Biochem.*, <u>XI</u>, 27 (1942).
78. *J. Am. Chem. Soc.*, <u>64</u>, 2716 (1942).
79. *J. Opt. Soc. Am.*, <u>32</u>, 635 (1942).
80. <u>The Structure and Properties of Synthetic Polymers</u>, <u>Nucleus</u>, (Dec., 1942).
81. *J. Am. Chem. Soc.*, <u>26</u>, 4 (1943), (with K. H. Sun).
82. *Ind. Eng. Chem.*, <u>35</u>, 216 (1943).
83. *Chem. Revs.*, <u>32</u>, 195 (1943).
84. *J. App. Phys.*, <u>14</u>, 246 (1943).
85. *J. Chem. Physics*, <u>11</u>, 412 (1943).
86. *J. Chem. Physics*, <u>11</u>, 419 (1943).
87. Thermodynamic Properties of Solutions of Long Chain Compounds. <u>Cellulose and Cellulose Derivatives</u>, edited by Emil Ott, Interscience Publishers, Inc., N.Y. (1943), pp 893-909.
88. <u>Cellulose and Cellulose Derivatives</u>, edited by Emil Ott, Interscience Publishers, Inc., N.Y. (1943), pp 943-955.
89. *Ind. Eng. Chem.*, <u>35</u>, 980 (1943). Reprinted in *Rub. Chem. and Tech.*, <u>17</u>, 38 (1944).
90. *Ann. N.Y. Acad. Sci.*, <u>44</u>, 431 (1943).
91. *Glass Ind.*, <u>24</u>, 472 (1943), (with K. H. Sun).
92. *Am. Scientist*, <u>31</u>, 338 (1943).
93. *J. Phys. Chem.*, <u>47</u>, 502 (1943).
94. *J. Am. Cer. Soc.*, (with K. H. Sun), <u>27</u>, 10 (1944).
95. *J. Am. Cer. Soc.*, (with K. H. Sun), <u>27</u>, 13 (1944).
96. *Rheology Journal - Letter to the Editor, Rheology Bulletin*, XIV, No. 4, p. 52, November 1943.
97. (With K. H. Sun and A. Silverman), *J. Am. Cer. Soc.*, <u>26</u>, 393 (1943).
98. Methods for the Analysis of Complex Molecular Structures with the Aid of X-Rays, <u>Colloid Chemistry: Theoretical and Applied</u>, Vol. V, edited by J. Alexander (1944).
99. The Vitreous State, <u>Colloid Chemistry: Theoretical and Applied</u>, Vol. V, edited by J. Alexander (1944). (With K. H. Sun and A. Silverman.)
100. X-Ray Diffraction and Its Applications. <u>Major Instruments of Science and Their Applications to Chemistry</u>, edited by R. E. Burk and O. Grummitt (1945).
101. *J. Am. Chem. Soc.*, <u>66</u>, 1991 (1944).

102. Photography of Crystal Structures, *Science,* <u>100</u>, 570 (1944).
103. Photography of Crystal Structures, *J. Chem. Phys.,* <u>12</u>, 520 (1944).
104. Photographic Fourier Synthesis, *Nature,* <u>155</u>, 18 (1945).
105. Comparison of the Structures of Stretched Linear Polymers, *J. Chem. Phys.,* <u>13</u>, 37 (1945). Reprinted in *Rub. Chem. and Tech.,* <u>18</u>, 763 (1945).
106. *Syn. Org. Chem.,* <u>17</u>, No. 1, (1945).
107. (With K. H. Sun), *J. Am. Chem. Soc.,* <u>28</u>, 149 (1945); *J. Soc. Glass Tech.,* <u>28</u>, 463-68 (1944).
108. *Polymer Bull.,* <u>1</u>, 25-30 (1945).
109. *Pictures of Organic Molecules from X-Ray Diffraction Data, Radiography and Clinical Photography,* <u>21</u>, 67 (1945).
110. (With K. H. Sun), *J. Amer. Ceramic Soc.,* <u>28</u>, 306-307 (1945); *J. Soc. Glass Tech.,* <u>29</u>, 192 (1945).
111. *J. Polymer Research,* <u>1</u>, 1 (1946).
112. (With K. H. Sun and R. M. Welch), *J. Amer. Ceramic Soc.,* <u>29</u>, 59 (1946); *J. Soc. Glass Tech.,* <u>30</u>, 333 (1946).
113. (With K. H. Sun), *J. Phys. Chem.,* <u>50</u>, 319 (1946); *J. Soc. Glass Tech.,* <u>30</u>, 318 (1946).
114. (With K. H. Sun), *J. Amer. Ceramic Soc.,* <u>29</u>, 232 (1946); *J. Soc. Glass Tech.,* <u>30</u>, 327 (1946).
115. (With K. H. Sun), *J. Phys. Coll. Chem.,* <u>51</u>, 438 (1947).
116.

117. Theorie des Proprietes Thermodynamiques des Solutions de Molecules en Chaine, *J. Chim. Phys.,* <u>44</u>, 9 (1947).
118. Theorie de la Haute Elasticite, *J. Chim. Phys.,* <u>44</u>, 99 (1947).
119. *J. Phys. Colloid. Chem.,* <u>52</u>, 248 (1948).
120. Fluophosphate Glass. (With K. H. Sun), U. S. Patent 2,481,700 (September 13, 1949).
121. Fluoride Glass. (With K. H. Sun), U. S. Patent 2,511,224 (June 13, 1950).
122. *J. Applied Phys.,* <u>21</u>, 518 (1950).
123. *Record of Chemical Progress,* <u>11</u>, 85 (1950).
124. "Transitions in Silver Halides." In <u>Phase Transformations in Solids</u> edited by Smoluchowski, Mayer and Weyl, Wiley (1951).
125. *J. Pol. Sci.,* <u>8</u>, 257 (1952).
126. *J. Am. Chem. Soc.,* <u>74</u>, 3963 (1952).
127. *J. Am. Chem. Soc.,* <u>74</u>, 3963 (1952)
128. *J. Am. Chem. Soc.,* <u>75</u>, 4123 (1953).
129. *J. Am. Chem. Soc.,* <u>75</u>, 4126 (1953).
130. *J. Am. Chem. Soc.,* <u>76</u>, 843 (1954).
131. *J. Am. Chem. Soc.,* <u>76</u>, 843 (1954).
132. *J. Am. Chem. Soc.,* <u>76</u>, 847 (1954).
133. *J. Chem. Phys.,* <u>22</u>, 1389 (1954).
134. *J. Am. Chem. Soc.,* <u>76</u>, 4045 (1954).
135. (With J. M. Stevels), *J. Am. Ceramic Soc.,* <u>37</u>, 474 (1954).
136. *J. Phys. Chem.,* <u>58</u>, 1141 (1954).

137. *Chem. & Engin. News,* <u>33</u>, 242 (1955).
138. *J. Polymer Sci.,* <u>16</u>, 209 (1955).
139. *J. Am. Ceramic Soc.,* <u>38</u>, 172 (1955).
140. *J. Am. Chem. Soc.,* <u>77</u>, 3928 (1955).
141. *Ann. Rev. Phys. Chem.,* <u>6</u>, 99 (1955).
142. *Bull. Chem. Soc. Japan,* <u>28</u>, 606-613 (1955).
143. *Bull. Chem. Soc. Japan,* <u>29</u>, 336-339 (1956).
144. "The Relation Between Structure and Properties of High Poly-
     mers." (In Japanese), <u>Proceedings Japan Chemical Fibers
     Association</u>, Dec. 16, 1955.
145. "Comments on the Future of High Polymer Science and Industry
     in Japan." (In Japanese), *High Polymers,* <u>5</u>, (1956).
146. "Thermodynamic Properties of High Polymer Solutions." (In
     English and Japanese), *High Polymers,* <u>5</u>, 494-498 (1956).
147. "The Structure of Glass." (In Japanese), *Journal of the
     Ceramic Association, Japan,* <u>64</u>, C390-392 (1956).
148. The World of the Future, English Mainichi (Osaka), (1956).
149. Valence, <u>The Encyclopedia of Chemistry</u>, edited by G. L. Clark,
     1957.
150. *Proc. Nat. Acad. Sci.,* <u>43</u>, 204 (1957).
151. *Proc. Nat. Acad. Sci.,* <u>43</u>, 209 (1957).
152. (With Y. Sakamoto), *J. Phys. Soc. Japan,* <u>12</u>, 241-251 (1957).
153. "Education and Research in Japan," *Genesee Valley Chemunica-
     tions,* <u>7</u>, No. 8, 6, 22, 24 (November 1956).
154. (With T. Abe), *J. Amer. Ceramic Soc.,* <u>40</u>, 287-292 (1957).
155. *J. Chem. Educ.,* <u>34</u>, 480-488 (1957).
156. "Some Aspects of the Structure of Proteins." <u>Chemistry of
     Proteins</u>, Vol. 5, edited by S. Akabori and S. Mizushima,
     Kyoritsu Syuppan Co., Ltd., Tokyo, 1957. (In Japanese)
157. *Hydrogen Bonds, Kagaku to Kogyo,* <u>10</u>, 606-616 (1957). (In
     Japanese, translated by H. Chihara.)
158. *Properties of High Polymers in Relation to their Structure,
     Kagaku,* <u>4</u>, 140 (1958).
159. <u>Physical Chemistry of High Polymers</u>, John Wiley & Sons, Inc.,
     New York, NY (1958).
160. *J. Polymer Sci.,* <u>30</u>, 5 (1958).
161. <u>Physical Chemistry of High Polymers</u>, (book in Japanese)
     Maruzen, Tokyo, Japan (1959).
162. *J. Polymer Sci.,* <u>43</u>, 473 (1960).
163. *J. Polymer Sci.,* <u>50</u>, 65 (1961).
164. <u>Actes Du Deuxienne Congres International de Catalyse</u>, Vol. 2,
     1399, Editions Technip, Paris (1961).
165. (With G. Natta, V. Desreux, and H. Mark), *J. Polymer Sci.,*
     <u>56</u>, 153 (1962).
166. *American Scientist,* <u>50</u>, 485 (1962); also (in Japanese) as a
     chapter in <u>S. Yamabe</u>, "Physical Chemistry of Drugs," Asakura,
     Tokyo, (1963).
167. "Some New Developments Concerned with the Structure of Colla-
     gen." In <u>Collagen</u>, edited by N. Ramanathan; Interscience
     Publishers (1962).

168. "High Polymers of the Future." Kurashiki Rayon Monthly Reports, November (1962), page 9 (in Japanese).

169. *J. Polymer Sci.*, Part C, 445 (1963).

170. (With G. Natta, V. Desreux, and H. Marks), *La Chimica e l'Industria*, 46, 536 (1964).

171. (With Y. Sakurada and W. R. Anderson,Jr.), *J. Phys. Chem.*, 68, 1934 (1964).

172. *J. Amer. Chem. Soc.*, 86, 3535 (1964).

173. "Coordination Polymers." Proc. 8th International Conference on Coordination Chemistry, V. Gutmann, Editor. Springer-Verlag, Wien/New York (1964), p. 253.

174. *Makromol. Chem.*, 87, 119 (1965).

175. (With G. Natta, V. Desreux, and H. Marks), *Makromol. Chem.*, 82, 1 (1965).

176. (With G. Natta, V. Desreux, and H. Marks), *Kobunshi*, 14, 708 (1965). (In Japanese.)

177. (With G. Natta, V. Desreux, and H. Marks), *Pure and Applied Chem.*, 12, 645 (1966).

178. *Makromol. Chem.*, 92, 260 (1966).

179. *Pure and Applied Chem.*, 12, 427 (1966). Also in Makromolecular Chemistry - 2, Butterworths, London, 1966.

180. "Some Theoretical Considerations Related to Defects in Glass." Symposium on Defects in Glass, Tokyo-Kyoto, 129 (1966).

181. "Theoretical Considerations," in Polymer Fractionation, M. J. R. Cantow, Ed., Academic Press, New York (1967). With H. Okamoto.

182. "Interactions Between Nonbonded Atoms." In Structural Chemistry and Molecular Biology, N. Davidson and A. Rich, Eds., Freeman, San Francisco, (1967).

183. *Pure and Applied Chem.*, 15, 369 (1967).

184. *J. Polymer Sci.*, (C) No. 23, 343 (1968). With K. Ohtsuka and S. Morimoto.

185. *J. Polymer Sci.*, (C) No. 23, 401 (1968). With Z. Reyes, M. Syz, and C. R. Russell.

186. *Macromolecules*, 1, 184 (1968).

187. *Inorg. Chem.*, 7, 2108 (1968).

188. Preface. (In English and Japanese). In An Introduction to Biothermodynamics (In Janapese), by S. Yamabe. Published by Nankodo, Tokyo (1968).

189. Discussion. In Proceedings of the Robert A. Welch Foundation Conferences on Chemical Research. X. Polymers, Houston, 1967.

190. *Polymer Preprints*, 10, 334 (1969).

191. *J. Paint Techn.*, 41, 509 (1969).

192. *J. Chem. Documentation*, 9, 230 (1969).

193. *J. Phys. Chem.*, 74, 371 (1970).

194. *Mitteilungsblatt Chemische Gesellschaft*, 17, 8 (1970).

195. *Revista de Plasticos Modernos*, 169, 553 (1970).

196. *J. Solid State Chemistry*, 2, 385 (1970). With R. A. Huggins.

197. *Acta Cryst.*, B26, 219 (1970).
198. "Discussion on Polymer Solutions." *Disc. Faraday Soc.*, No. 49, 164, 175 (1970).
199. *Product Research and Development*, 9, 121 (1970).
200. "50 Jahre Theorie der Wasserstoffbruckenbindung." *Angew. Chem.*, 83, 163 (1971).
201. *Inorg. Chem.*, 10, 791 (1971).
202. *J. Phys. Chem.*, 75, 1255 (1971).
203. *Polymer*, 12, 357 (1971).
204. *Macromol.*, 4, 274 (1971).
205. "50 Years of Hydrogen Bond Theory." *Angewandte Chem.*, Internat. Ed., 10, 147 (1971).
206. *J. Polymer Sci.*, C33, 55 (1971).
207. Kinetics and Mechanism of Polyreactions, IUPAC, Vol. VI., Plenary and Main Lectures. Akademiai Kiado, Budapest (1971).
208. Proc. 1st International Conference on Calorimetry and Thermodynamics, Warsaw, 1969. Polish Scientific Publishers, Warsaw (1971).
209. Amorphous Materials, edited by R. W. Douglas and P. Ellis, John Wiley & Sons, Ltd., Chichester, England (1972).
210. *J. Paint Technology*, 44, 55 (1972).
211. *Pure and Applied Chem.*, 31, 245 (1972).
212. Ansprache (at the Ceremony for the award of the Honorary Doctor of Science degree). *Mitteilungsblatt Technische Universitat Clausthal*, 27, 22 (1972).
213. *J. Ceramic Society, Japan*, 80, 473 (1972).
214. *Br. Polym. J.*, 4, 465 (1972).
215. *Rev. Mod. Plast.*, 201, 387 (1973).
216. *Polymer J.*, 4, 502 (1973).
217. *Polymer J.*, 4, 511 (1973).
218. *Kolloid-Z U. Z. Polymere*, 251, 449 (1973).
219. *J. Polymer Sci.: Symposium No. 42*, 1171 (1973).
220. *Wissenschaftliche Zeitschrift der T. H. Leuna-Merseburg*, 17, 125 (1975).
221. Proc. Int. Conf. on Colloid and Surface Science, (E. Wolfram, Ed.), Akad. Kiado, Budapest, 1975, p. 3.
222. Molecular and Intermolecular Properties from Excess Enthalpies. Preprints, Fourth International Conference on Chemical Thermodynamics, Montpellier, France, 1975, VI/12.
223. G. Gabrielli, E. Ferroni, and M. L. Huggins, *Progr. Colloid and Polymer Science*, 58, 201 (1975).
224. *J. Phys. Chem.*, 80, 1317 (1976).
225. "Some Horizons in Colloid Science." *Progr. Colloid Polymer Science*, (in press).
226. The Structure of Alpha Keratin. Schriftenreihe Deutsches Wollforschungsinstitut an der Technischen Hochschule Aachen (in press).
227. Learning about Polymers from Mother Nature (submitted for publication).

L. Mandelkern

Studies of crystalline polymers predate the recognition and acceptance of polymers as molecules of very high molecular weight[1]. These very early works were primarily concerned with the determination of crystal structures and unit cell dimensions by the conventional methods of x-ray diffraction. These investigators were confronted with the dilemma as to whether it was necessary for the complete molecule to be located within the unit cell. The resolution of this problem was one of the major steps which lead to the adoption of the macromolecular hypothesis[1]. In the 1940's there were two major advances in other aspects of this general problem. Wood and Bekkedahl[2] reported the results of an experimental study of the crystallization kinetics and melting of natural rubber. Although these observations could not be interpreted until much later, they forcibly demonstrated the fact that crystalline polymers could be quantitatively studied using methods which were well known for low molecular weight crystalline substances. Flory[3] presented thermodynamic theory of the fusion of polymers in 1949 which has had many major and far reaching consequences. Although not always directly recognized as such, these theoretical considerations have continued to serve as a base for much of the research in this field today.

The past twenty-five years, coincident with the initiation and growth of the Division of Polymer Chemistry, has witnessed an unprecedented level of research activity in the field of crystalline polymers. There are many reasons for this development. Foremost, it is the large number of basic scientific problems, involving different disciplines and methods of approach, which are generated by the three dimensional organization (and melting) of a collection of long chain molecules. Crystallinity and crystallite orientation

113

exert a major influence on the physical and mechanical   properties
of a polymer system and thus dictate the end-use of a particular
system.   Research has therefore also been greatly stimulated by
technological demands.   In order to discuss the present status of
this field and to be able to make some reasonable projections for
the future it seems logical and advisable to first examine what has
become well established and deep-rooted in the past so that we can
understand the present.

As has been indicated, Flory[3] developed a statistical thermodynamic theory for the fusion of crystalline polymers which was
verified by a variety of experimental studies [4-6].   The underlying
conclusion was that the crystallization-melting process involving
long chain molecules was a first-order phase transition, albeit a
diffuse one.   A well defined equilibrium melting temperature exists
which systematically depends on added diluent concentration,[3] copolymer sequence distribution and composition [3,7], molecular weight
with the most probable distribution[3] and chain length when uniform[8].
The amorphous and crystalline phases can coexist in equilibrium below the melting temperature[3,9].   The concept of a first-order phase
transition and an equilibrium melting temperature is a requirement
for the application of nucleation theory which as proven to be such
an important and cogent idea in the understanding of crystallization kinetics and the resulting properties[6,10].

In these formative, but important stages of development and understanding, it was also shown that the overall crystallization
process from the pure melt[6,10], from polymer-diluent mixtures[6,11]
and more recently from very dilute solution[12,13] follows the general mathematical formulation for the kinetics of phase changes as
developed by Avrami for metals[14].   A very strong negative temperature coefficient of the crystallization rate is invariably observed in the vicinity of the melting temperature[6].   This phenomena is
recognized as a general characteristic of nucleation control of the
crystallization process.   Unfortunately, although it is extremely
important, the general nature of this conclusion precludes the establishment of a unique molecular description of the nucleation
process and the nature of the chain disposition within the nucleus.
This restriction has not been completely recognized in variou arguments that have been put forth for specific kinds of molecular nucleation acts.

Some very important principles, which are fundamental to the
comprehension of properties of crystalline polymers were established at this point.   Despite the apparent molecular complexities,
polymer crystallization follows the same scientific principles already established for other crystalline substances.   However, many
important questions, unique to polymers, and related mainly to macroscopic properties remained unanswered at that time and some still

remain so. The reason is that although the correct physico-chemical framework was established real crystallization processes, involving polymers, yield non-equilibrium systems, because the crystallization must be invariably conducted at temperatures well below the equilibrium melting temperature. Thus crystallization kinetics and mechanisms determine properties through the morphological forms that evolve and the accompanying chain disposition. Thus studies of morphology and crystallization mechanisms became an endeavor in the studies of crystalline polymers.

Very early light miscroscopy studies[15], with branched polyethylene (low density polyethylene), indicated a spherulitic structure similar to that found in many other polycrystalline organic and inorganic stbstances[6]. These represent a definite organization and orientation of crystallites relative to one another. Spherulitic structures were subsequently observed in such a wide variety of homopolymers crystallized in the bulk as to appear to be the universal mode of polymer crystallization. More recent work[16, 17, 18] (see below) has tempered this conclusion with restrictions of molecular weight, molecular weight distribution and crystallization temperature.

In the late 1950's a major advance in the understanding of morphology was made with the virtually simultaneous reports from several different laboratories[19, 22] observations which confirmed a long neglected paper by Storks[23]. It was observed that homopolymers precipitated from dilute solution in the form of platelets or lamellar-like cyrstallites. Electron micrographs of such lath-like crystals have now been widely observed, are well known, and can be considered to be a universal mode of homopolymer crystallization[24]. Typically such crystals are several microns in lateral dimensions and the order of 100 Å thick. Most important is the fact that the chain axis is perpendicular to the wide face of the crystal. Since the platelet thickness is no more than the order of a few hundred angstroms and this crystal habit is observed for very high molecular weight chains, it is then necessary for a molecule to traverse a crystallite many times. This reentrant requirement demands some type of "folded chain". However, the detailed nature of the reentry, which governs the interfacial structure, is not at all obvious nor can it be obtained solely from an examination of electron micrographs. Despite this well grounded concern, it was postulated, based primarily on the external shape of the platelet crystallites and the apparent smoothness of the basal planes, that the chains were regularly folded. This concept, which enjoyed great popularity in its time, required that crystallization be complete except for the number of chain elements that are required to make the hairpin connections between adjacent crystalline sequences. It was thus proposed that a regularly folded, pleated, smooth 001 interface is

formed[21,25,28]. Various theroretical basis for these far sweeping
conclusions were proposed,[29,35] which unfortunately either did not
really direct themselves to the issue at hand or were not subsequently substantiated[36].

Lamella-like crystallites were also revealed to be the major
morphological entity of homopolymers crystallized from the melt.
The chain orientation is the same as found in solution crystals
and this typical crystal habit explains the chain orientation
found within spherulites. However, the nature of the concerted
action of a collection of lamella to form a spherulite still remains to be explained. These initial studies, which were conducted with rapidly crystallized samples, gave crystallite thickness which were of the same order as solution crystallized material.
Hence a correspondence between the two cases was naturally immediately made. Thus, when it was assumed that the surface of solution
formed crystals consisted of regularly folded chains the same interfacial structure was assigned to bulk crystallized polymers. By
necessity, therefore, all the chain units were assigned to the
interior of crystallites or to the interface. Chain units in non-
ordered conformations connecting crystallites (amorphous regions)
did not exist in this view. The well known major deviations in
properties from that expected of a macroscopic crystal were attributed to contributions from the interface as well as from defects
within the interior of the crystallites. It was proposed that a
crystalline polymer should be viewed as consisting of disordered
material or defects which are embedded within the crystalline
matrix[27]. These conclusions, had an obvious and very important
implication to the interpretation of the properties of crystalline
polymers. They were based, however, on gross or non-molecular
morphological observations. The unprecedented concentration of
defects that would be required to explain the properties were
tacitly ignored. Despite these misconceptions, the experimental
discovery of lamella-like crystallites and the accompanying chain
orientation represented a major advance in understanding the
structure of crystalline polymers. They can furthermore, properly
serve as a base for further study of the molecular structure and
chain conformations when it is recognized that the lamella
crystallites do not require a smooth, regularly chain folded interface. The lamella, as observed under the electron microscope, are
compatable with other types of interfacial structures.

A fruitful approach, which has led to a better understanding
of these structures, has been obtained through the systematic
study of properties. Wide spread studies, by many investigators,
of many different properties of solution formed crystals has lead
to the conclusion that the interface can not be smooth or regularly structured[24,36,39]. A disordered, amorphous overlayer is
clearly required by thermodynamics, spectroscopic, physical and

mechanical properties. The detailed nature of the chain reentry is still a matter of some disagreement[38],[39]. However, the preponderance of the evidence favors the random connection of crystalline sequences by loops of random length, as originally proposed by Flory[40] and by Fischer[41] rather than adjacent reentry with disordered loops[42],[43].

The removal of the stricture of a regular folded interface allows for an objective analysis to be made of the structure of bulk crystallized polymers. Consistent with the gross morphological observations these concepts can be developed from a study of properties, which have been found to be dependent on the molecular weight (when fractions are used) and the isothermal crystallization temperature[44],[45],[46]. From an analysis of many different kinds of properties, controlled in this manner, a model of a primitive crystallite can be developed which represents a level of order above that of the unit cell[36-45]. There are in general three major regions associated with such a crystallite. There is an interior core, or crystalline region, comprised of chain sequences in ordered conformation giving rise to the lamella habit. The length of the ordered sequences, or thickness of the crystalline regions, are determined by crystallization temperature and molecular weight through nucleation control[44],[46]. The level of imperfections within the crystalline region is found to be about the same as in comparable monomeric analogues[36]. There is, therefore, no basis to postulate unduly large concentrations of undefined defects to explain properties. Associated with this crystalline portion is an interfacial region or zone. In contrast to monomeric systems it is not sharply defined but is rather diffuse and can be many units thick. Interfacial free energies associated with this basal plane are relatively high when compared to low molecular weight substances and depend a great deal on the relation between the crystal thickness and extended chain length[47],[48]. Chain units in non-ordered conformations connect the crystallites and form the interzonal or amorphous region. The properties of the chain in this region are very similar, if not identical, with the completely molten polymer at the same temperature[36],[37]. Properties depend on the relative proportion of these regions which can be systematically varied by control of molecular weight and crystalization temperature.

This working model of a crystallite represents the first rudimentary steps in what can be thought of as molecular morphology. In this endeavor the gross morphology, as is observed by microscopic techniques, is further refined in terms of molecular structure. The chain conformation and disposition within and without the crystallites, and the nature and concentration of defected structures need to be defined. What has already been accomplished represents only the genesis of what is ultimately required to describe molecular structure and to relate it, in turn, to macroscopic properties. Within the framework that has been established for the crystallite structure a more detailed description of the interfacial region is required for both solution and bulk crystallized samples. The key matters here are the specific nature of the re-entry process and the thickness, in terms of chain units, of the disordered zone. Refinements are also required in the structural description of the interzonal region. The main point is whether there are any perturbations in the conformation from that of the pure melt and the influence, if any, of the supermolecular structure. Although these are fairly subtle questions they are important in interpreting properties and could possibly affect the location of the glass temperature in semi-crystalline polymers.

Several relatively new techniques show promise in helping to resolve small-angle neutron scattering studies of mixtures of deuterated and hydrogenated polyethylene demonstrate the capacity of describing the non-crystalline regions through measurements of the radius of gyration[49]. One of the main problems that needs to be solved to make this method effective, is preparing samples of uniform composition[50]. Studies of the carbon-13 nmr relaxation parameters[51] of partially crystalline linear polyethylene[52] and cis poly-isoprene[53] have detected differences in amorphous structure with major changes in morphological form.

As has been indicated, the lamella crystallites can be further organized into higher levels of morphology or supermolecular structure. Spherulitic structures are well known examples and are widely observed in crystalline polymers. Until recently they had been thought to be the universal mode of homopolymer crystallization. However, it has been recently found from small angle light scattering studies[54], with molecular weight fractions of linear polyethylene, that different morphologies are developed depending on the molecular weight[16]. In this first study mainly rapidly cooled samples were used. Hence, the effect of isothermal crystallization at low undercooling was not assessed. However, restrictions on spherulitic formation were clearly established so that this morphological form can no longer be considered to be a universal mode of homopolymer crystallization. For linear polyethylene well-defined spherulites were found in the molecular weight range 1.2×10^4 to 9×10^5. However, for $M = 6 \times 10^5$ and 9×10^5 the light scattering patterns indicate imperfections of either shape or internal structure of the spherulites. The spherulite morphology is not maintained for $M > 9 \times 10^5$. A rod like morphology is found for $M = 1-2 \times 10^6$. Above this molecular weight the superstructure is ill-defined and disordered[16]. It has now been found that spherulites will not form in this molecular weight range at any crystallization temperature. It has also been observed that samples of the same density

can be prepared which, depending on the molecular weight, can or
can not be spherulitic.

More detailed studies of the supermolecular structure under
isothermal crystallization conditions are clearly required. Low
angle light scattering appears to be the most objective way of ac-
complishing this goal with selected complementary light microscopy
observations. Light microscopy studies of a polyethylene fraction
$M = 3 \times 10^4$ indicates a critical crystallization temperature below
which spherulites are formed and above which they are not[18]. A
similar result has been found for $M = 6 \times 10^5$ by light scattering
measurement[17]. Changes in the growth mechanism are indicated and
the temperature at which they take place can be correlated with
the temperature coefficient of the overall crystallization kinetics[55]. These preliminary results substantiate the need for further
studies to delineate the conditions for the formation of the different kinds of supermolecular structures that have been observed.
Moreover, the relationship of properties to these structures needs
to be developed.

Since spherulites are not formed from high molecular weight
chains it is of interest to learn whether spherulitic development
is impeded by the addition of such species to a spherulite forming
system. In an exploratory study[17] a surprisingly small amount of
high molecular weight material (10-20% by weight) reduces the
spherulite size and eventually cause their disappearance in a
molecular weight fraction in which well developed spherulites are
usually observed. The effect of mixtures, and polydispersity in
general, needs to be studied in more detail as we now have the potential for major control of supermolecular structure and consequently properties.

With these recent developments in molecular morphology, supermolecular structure and properties attention needs to be redirect-
ed to a molecular understanding of the crystallization mechanism.
From this will stem naturally our understanding of the morphology
and properties. A well developed phenomenological base has already been established[6]. Unfortunately, it does not lead to a
unique molecular interpretation. The development of molecular
models for the kinetic processes are in the early stages of development[56] and must await the tests of time and experiment to attest
their validity. We must remember that the basic reason for the
formation of the lamella-like crystallites has not as yet been
firmly established[40,46].

This report and the work described therein from out laboratory
was supported by Grant DMR72-03052 from the National Science Foundation.

REFERENCES

1. P. J. Flory, Principles of Polymer Chemistry, Cornell Press, 1953, p. 22.
2. L. A. Wood and N. Bekkedahol, J. Appl. Phys., 17, 362 (1946).
3. P. J. Flory, J. Chem. Phys., 17, 223 (1949).
4. L. Mandelkern, Chemical Reviews, 56, 903 (1956).
5. L. Mandelkern, Rubber Chem. Tech., 32, 1392 (1959).
6. L. Mandelkern, Crystallization of Polymers, McGraw-Hill (1964).
7. P. J. Flory, Trans. Farad. Soc., 51, 848 (1955).
8. P. J. Flory and A. Vrij, J. Amer. Chem. Soc., 85, 3548 (1963).
9. R. Chiang and P. J. Flory, J. Amer. Chem. Soc., 83, 2857 (1961).
10. L. Mandelkern, F. A. Quinn, Jr., and P. J. Flory, J. Appl. Phys. 25, 830 (1954).
11. L. Mandelkern, J. Appl. Phys., 26, 443 (1955).
12. C. Devoy, L. Mandelkern, and L. Bourland, J. Polym. Sci., A-2, 7, 869 (1970).
13. E. Riande and J. M. G. Fatou, Polymer, 17, 99 (1976).
14. M. Avrami, J. Chem. Phys., 7, 1103 (1939); ibid. 8, 212 (1940).
15. C. W. Bunn and T. C. Alcock, Trans. Farad. Soc., 41, 317 (1945).
16. S. Go, L. Mandelkern, R. Prud'homme, and R. Stein, J. Polym. Sci., Polym. Phys. Ed., 12, 1485 (1974).
17. S. Go, L. Mandelkern, D. Pfeiffer, and R. Stein, in preparation.
18. J. D. Hoffman, G. T. Davis, and J. I. Lauritzer, Jr. (1976), in Treatise on Solid State Chemistry, D. Hanney, ed., Plenum Press, Vol. 3.
19. R. Jaccodine, Nature, 176, 301 (1955).
20. P. H. Till, J. Polymer Sci., 24, 301 (1957).
21. A. Keller, Phil. Mag., (8) 2, 1171 (1957).
22. E. W. Fischer, Z. Naturforsch, 12a, 753 (1957).
23. K. H. Storks, J. Amer. Chem. Soc., 60, 1753 (1938).
24. L. Mandelkern, (1970), in Progress in Polymer Science, A. D. Jenkin, ed., Pergamon Press, Vol. 2, p. 165.
25. A. Keller, Makromol. Chem., 34, 1 (1959).
26. A. Keller, Polymer, 3, 393 (1962).
27. P. H. Lindenmeyer, Science, 147, 1256 (1956).
28. J. D. Hoffman, SPE (Soc. Plast. Eng.) Trans., 4, 315 (1964).
29. A. Peterlin, J. Appl. Phys., 31, 1934 (1960).
30. A. Peterlin and E. W. Fischer, Z. Phys., 159, 272 (1960).
31. A. Peterlin, E. W. Fischer, and C. Reinhold, J. Chem. Phys., 37, 1403 (1962).
32. A. Peterlin and C. Reinhold, J. Polym. Sci., Part-A 3, 2801 (1965).
33. F. P. Price, J. Chem. Phys., 35, 1884 (1961).
34. J. I. Lauritzen, Jr. and J. D. Hoffman, J. Res. Nat. Bur. Stand., Sect. A, 64A (1960).
35. F. C. Frank and M. Tosi, Proc. Roy. Soc. Ser., A, A263, 323 (1961).

36. L. Mandelkern, (1975), in Characterization of Materials in Research: Ceramics and Polymers, Proceedings of the 20th Sagamore Army Materials Research Conference, Syracyse University Press.

37. L. Mandelkern, *J. Polym. Sci.*, C50, 457 (1975).

38. L. Mandelkern, *Acct. of Chem. Research*, 9, 81 (1976).

39. L. Mandelkern, *Annual Rev. of Materials Sci.*, Vol. 6, (1976).

40. P. J. Flory, *J. Amer. Chem. Soc.*, 84, 2857 (1962).

41. E. W. Fischer and G. Schmidt, *Angew Chem.*, 74, 551 (1962).

42. H. G. Zachmann and A. Peterlin, *J. Macromol Sci.-Phys.* 3, 495 (1969).

43. M. I. Bank and S. Krimm, *J. Polym. Sci.*, A-2 7, 1785 (1969).

44. J. G. Fatou and L. Mandelkern, *J. Phys. Chem.*, 69, 417 (1965).

45. L. Mandelkern, *J. Phys. Chem.*, 75, 3920 (1971).

46. L. Mandelkern, *Polym. Sci., Eng.*, 7, 232 (1967).

47. L. Mandelkern, J. M. Price, M. Gopalan and J. G. Fatou, *J. Polym. Sci.*, Part A 4, 385 (1966).

48. J. M. Schultz, W. H. Robinson and G. M. Pound, *J. Polym. Sci.*, A-2, 5, 511 (1967).

49. J. Schelten, G. D. Wignall and D. G. H. Ballard, *Polymer*, 15, 682 (1974); G. D. Wignall, private communication.

50. F. Stehling, E. Ergöz and L. Mandelkern, *Macromolecules*, 4, 672 (1971).

51. J. Schaefer, (1974), in Topics in Carbon-13 NMR Spectroscopy, G. C. Levy, ed., Wiley-Interscience, New York, Vol. 1, p. 149.

52. R. A. Komoroski, J. Maxfield, F. Sakaguchi and L. Mandelkern, submitted to *Macromolecules*.

53. R. A. Komoroski, J. Maxfield and L. Mandelkern, in preparation.

54. R. S. Stein (1964), in New Methods of Polymer Characterization, B. Ke, ed., Wiley-Interscience, New York.

55. E. Ergöz, J. G. Fatou and L. Mandelkern, *Macromolecules*, 5, 147 (1972).

56. I. C. Sanchez, *J. Macromol. Sci. -Revs. Macromol. Chem.*, C10, 113 (1974).

H. F. Mark

When I arrived from Canada in Brooklyn in September 1940, I was attached to the Shellac Bureau whose Director, Professor William Howlett Gardner, had established a National Testing Laboratory for this important natural resin and, at the same time, was carrying out very interesting studies on its chemical composition and structure. We soon agreed that, in view of the World Situation considerable interest and effort should be devoted to synthetic coatings and film formers and that a few graduate students should start work in this direction. When we submitted to Raymond E. Kirk, Head of the Chemistry Department, a proposal for space, equipment and personnel, he immediately agreed and has always, up to his premature death in 1957, been a benevolent and deeply understanding champion of all polymer work at the Polytechnic Institute of Brooklyn. The expereiences collected at the I. G. Farben Laboratory in Ludwigshafen (1926-1932) at the University of Vienna (1932-1938) and at the Canadian International Paper Company in Hawkesbury, Ontario (1938-1940) made it clear that systematic progress could only be espected if organic chemists, physical chemists and at least one physicist could be assembled into a team to conduct research from the monomer or monomers through the polymerization process and the characterization of the resulting materials to the structure and properties of such final products as coatings, films and fibers.

Since Professor Gardner soon went to Washington, D.C. on official mission and no young organic chemist was immediately available we started to concentrate on the mechanism of polymerization processes, on molecular weight and molecular weight distribution and on the solution properties of polymers. Much work had been

done in Europe in the late 1930's on suspension - and emulsion
polymerization, particularly on its initiation and regulation by
redox systems. Although a few industrial laboratories were aware
of these efforts, their results were not generally known and it
seemed appropriate to carry out a few additional clarifying experiments and to publish a comprehensive review article. As the number
of associates and graduate students increased the name of the
Shellac Bureau was changed into Polymer Research Institute and
PhD degrees in Polymer Chemistry were authorized by the Dean of the
Graduate School and the faculty. At this time the volume of literature emerging from universities, industry and Government Laboratories increased so rapidly that it was apparently worthwhile
to think about some adequate vehicles for the concentration of information in the polymer field and, in fact, Drrs. E. S. Proskauer
and M. Dekker of Interscience Publishers consented to initiate in
1940 the publication of a series of Monographs on "High Polymers"[1]
with H. Mark and G. S. Whitby as editors and to start publishing
in 1946 the Journal of Polymer Science[2].

Most scientific and technical work in the polymer field was
carried out during the early forties in industry. All larger
chemical companies maintained research and development laboratories
which worked on plastics, adhesives, coatings, film and fibers and,
under the impact of the developing rubber shortage and the Government sponsored synthetic rubber programs, most rubber companies did
research on polymerization and polymer rheology; even a few larger
oil companies like Esso, Shell and Phillips demonstrated considerable interest in the synthesis, characterization and application
of polyolefins, vinyls and polydienes. It is probably correct to
estimate that 2500 to 3000 chemists and engineers were occupied at
that time in American industry with studies in the general field
of polymers. In comparison there was little activity in academic
institutions although there existed a few individual groups who
were carrying out very important pioneering work in various directions such as C. S. Marvel and F. T. Wall in Illinois, P. Debye
and P. J. Flory in Cornell, W. H. Stockmayer at M.I.T., R. M. Fuoss
in Yale, A. V. Tobolsky in Princeton, F. d'Alelio and E. Guth in
Notre Dame and Milton Harris and his group at the National Bureau
of Standards. Several approaches were taken at that time to ac-
celerate the flow of new fundamental information from academia to
industry and that of urgent practical problems from industrial laboratories to university research centers. One was the so-called
"Gibson Island Research Conferences" which was founded in the late
1930's by Professor Neil Gordon of Johns Hopkins University in
Baltimore and represented an ideal trading post for scientific and
technical information and discussion in an extremely attractive,
informal and relaxing atmosphere. Particularly the "Polymer" and
"Fiber" weeks generated tremendous attraction and were always
strongly oversubscribed. It is well known that in the course of

the years these vacation gatherings expanded into the enormous
framework of the "Gordon Research Conferences" which - now number-
ing more than 60 every year - are internationally recognized as the
best and most successful means to exchange ideas, expereiences and
concepts in the entire field of natural sciences. Universities,
governmental agencies and industry participate on equal terms and
everybody agrees that the informal, congenial atmosphere of the
Gordon Conferences was not only an important instrument to create
intellectual progress but also a valuable factor in the promotion
of international relationship between scientists of all branches.

Our Institute in Brooklyn is surrounded by many other schools
of higher learning and also by numerous industrial organizations
which maintain smaller or larger research and development laboratories. It soon became obvious that some special means should be
created to advance the scientific and technical interchange with
these institutions in the polymer field. This was accomplished in
two ways. First we established a series of regular seminars and
symposia every Saturday from October to May, where certain, particularly actual and timely subjects were presented and discussed.
There were usually between two and four presentations and there
was ample time for discussion and, if necessary, for additional
short contributions. As the years went by these "Saturday Symposia"
became a widely known and well frequented institution which started
to assume even international recognition when many of our distinguished foreign visitors acted as chairmen, speakers or discussion
leaders. The second vehicle for a closer cooperation between our
Institute and industry were our "Special Summer Courses" in which
a limited number (10-20) of visiting scientists remained at the
Institute for the period of one to two weeks and were given detailed theoretical and experimental instruction in a special field such
as "X-ray Diffraction", "Molecular Weight Determination", "Suspension Polymerization" and others. The first "X-ray Course" was
given by I. Fankuchen and H. S. Kaufman; the first "Molecular
Weight Course" by T. Alfrey, P. M. Doty and B. H. Zimm. For the
School we started to develop a comprehensive Graduate Curriculum
for Polymer Scientists with such courses as: "Mechanism of Polymerization Processes", "Solution Properties of Polymers", "Structure
and Properties of Polymers in the Solid State", "Suspension and
Emulsion Polymerization", and others. Scientifically our work was
essentially a continuation of what had been the program in Vienna
and in Hawkesbury with several improvements, refinements and extensions. A book was written with A. V. Tobolsky[3] on the "Principles of the Physical Chemistry of Polymers" and another one with
R. Raff[4] on "Polymerization Reactions", a detailed systematic
study of the influence of concentration, solvent, catalysts and
temperature on vinyl type addition polymerization was made, fractionation techniques for cellulose acetate and polystyrene were
improved, special attention was given to experimental perfection

in osmometry, viscometry and x-ray analysis of partially crystalline systems.

But there were also a number of approaches which were new and original. One of them was the use of radioactive bromine as a label for certain endgroups in vinyl polymerization. Together with fractionation, osmotic and visometric measurements (and later particularly with light scattering data) its results allowed interesting conclusions on the termination and branching processes and on the infrastructure of addition polymers.

Another interesting innovation was the use of hydrogen-deuterium exchange rate measurements[5] for the establishment of the accessibility of the different hydroxyl groups in a polymeric system. In swollen samples of partially crystallized cellulose such measurements allow a new approach to the "degree of crystallinity" via the quantitative establishment of non reactive i.e. nonaccesible hydroxyl groups. This method has since been greatly improved and is now much used in such fibers as nylon, polyvinylalcohol, cellulosics and in many proteinic systems.

Still another new approach[6] was the combined use of fractionation, osmometry and viscometry to study the influence of the polymerization conditions (temperature, extent of polymerization, solvent) on the infrastructure of addition polymers; it was originally exemplified with three polystyrene samples prepared at different temperatures but has since been used in refined form and with such improved fractionation techniques (GPC) to study such problems as branching, stereoregulation and grafting.

Another little trick which became very popular and provided for much interesting information was the use of a somewhat adjusted Linderstroem-Lang gradient tube for the study of semicrystalline polymeric systems such as fibers at different degrees of orientation and crystallinity[7]. It permitted, for instance to find a surprisingly large structural difference in small fiber fragments cut at the inside and outside of a bundle at the same position of its length apparently as a result of different tension and different rates of cooling.

After a few years our groups had grown to a substantial research force consisting of permanent staff members and of adjunct and visiting professors and comprising specialists in Biochemistry like Felix Bergman and K. G. Stern, organic chemistry like E. D. Bergmann, C. G. Overberger, C. C. Price and W. P. Hohenstein, physical chemistry like Turner like Turner Alfrey, P. M. Doty, F. R. Eirich, G. Goldfinger, A. V. Tobolsky, G. Oster, R. B. Mesrobian and B. H. Zimm and physicists like R. Brill, I. Fankuchen, E. Montroll, H. Riseman and R. Simha.

A few research areas of wider scope were selected for longer range and somewhat sustained efforts. One was the rate of di- fusion of gases and vapors through solid polymers. Besides the chemical composition of the interacting materials this phenomenon is influenced by many other factors and gave attractive opportun- ities to study the consequences of orientation, crystallinity, plasticizer content and, most interestingly, temperature. Studies[8] of this last effect led to the discovery of the "so-called Second Order Transition Point", today well known and much better under- stood as Tg. It also stimulated broader research on the mechanical properties in general with special emphasis on creep data of fibers and films on their quantitative interpretation with spring-dash- pot models. A unified aspect of fibers, plastics and rubbers was presented through the cooperation of three effects: the confor- mational flexibility of the individual chains, their infrastructure and the resulting capacity for regular lateral packing (crystalli- zation) and the magnitude of their intermolecular attraction. A series of individual publications culminated in Turner Alfrey's book on "The Mechanical Behavior of Highpolymers" which was the first comprehensive treatise of this field and for many years re- mained a classic in content and scope.[9]

In the days of Buna S and Buna N in Ludwigshafen the phenomenon of copolymerization received its first pioneering marks and, in the following years was studied with the accumulation of more and more material which, however, remained essentially qualitative and em- pirical. While we carried out numerous quantitative measurements on free radical homopolymerization it was evident that equivalent sets of tests should be made on copolymerization. Two questions[10] presented themselves for an experimental and theoretical investi- gation: copolymer composition and copolymerization rates in com- parison to the homopolymerization rates of the components. On the basis of existing data in the literature and of additional tests carried out by several of our graduate students, Alfrey and Gold- finger, developed the well known copolymerization composition equa- tions which permits to predict the composition of a copolymer from the composition of the monomer feed with the aid of two empirical parameters derived from low conversion runs over a range of feed compositions. Almost at the same time and independently several important articles were published by O. Lewis, F. R. Mayo and C. Walling which contributed a wealth of interesting experimental data and also culminated in the derivation of the (same) copolymer composition equation. All available material was subsequently used by Turner Alfrey and C. C. Price to clarify the fundamental back- ground for the perplexing variety of copolymerization conditions through the cooperation of two independent characteristics of each monomer - the electrostatic charge "c" and the resonance stabili- zation "Q" of its radical form, additional work on copolymerization rates and on properties led to the idea that it might be timely to

publish a comprehensive treatise of the entire field. This was
done in Volume VIII of the Polymer Series as "Copolymerization"
by T. Alfrey, J. J. Bohrer, and H. Mark.[11]

The advanced understanding of the various steps of polymerization reactions which was gradually reached in these years stimulated
the preparation of more complicated structures, which would contain
branched and/or segmented chains. Several isolated attempts had
been made in various laboratories without establishing any proof
for the structure of the resulting products. In 1948, T. Alfrey
initiated a systematic experimental study along these lines which
was later expanded by R. B. Mesrobian and E. H. Immergut. In 1952,
a comprehensive article was published describing several of the
useful methods to prepare these more complicated macromolecules
which were called "graft" and "block" copolymers[12] and which
displayed a number of unusual and attractive properties. Since
then, these more sophisticated synthetic manipulations have resulted
in many important new plastics, rubbers, and fibers and are now
summarized in several informatives and extensive monographs. At
that time -- beginning of the 1950's -- Polymer Science had become
a sufficiently large and important branch of chemistry to be
recognized by a special Division of Polymer Chemistry. In fact,
in 1948, a group of progressive pioneers in universities and
industry filed a petition with the American Chemical Society to
establish such a Division and, after a probation period of three
years, the Division of Polymer Chemistry was founded. Since then
it reflects in general and in details all important events in the
field and today with a membership of more than is one of the
larger Divisions of our Society.

Soon after the war it appeared appropriate to establish
Polymer Science as a significant and promising branch of Organic
Chemistry. This was first done in FAO through the formation of a
"Technical Committee for Wood Chemistry" and later by the foundation
of a "Macromolecular Commission" in the International Union of Pure
and Applied Chemistry. These extramural activities together with
frequent and necessary personal contacts with scientists in Europe,
Israel, Japan, and the Soviet Union took so much of my time that
Professor C. G. Overberger assumed the actual leadership of the
Polymer Research Institute in Brooklyn and maintained it even when
he was made -- later -- Head of the Chemistry Department and --
still later -- Dean of Science. His capacity to organize his time
and work was -- and still is -- so exceptional that he can maintain
creative and original research efforts in spite of time consuming
administrative responsibilities. Throughout the years, it was
customary that the Senior Professors would pursue their own R&D
projects independently and with a number of associates and co-
workers of their own choice: Turner Alfrey on Mechanical Behavior,
R. B. Mesrobian on Graftcopolymerization, Murray Goodman on
Stereochemistry of Polymers, F. R. Eirich on Rheology, G. Oster on

Photochemistry of Polymers, H. P. Gregor on Polyelectrolytes,
H. Morawetz on Solid State Polymerization, and C. G. Overberger on
a broad front of Organic Chemistry of monomers, catalysts, and
polymers.

Our Institute was only a small part of the fundamental and
practical polymer work which was going on in the USA in
Universities and Industry but we always attempted to emphasize
important results of other groups by Seminars and Symposia or by
exchanging scholars for shorter or longer periods. The same
tendency existed towards exceptional accomplishments abroad:
polycarbonate and polyurethane chemistry was introduced by special
lectures or seminars and -- later -- made even more available by
several volumes of the Polymer Series. The academic events during
Professor Natta's first visit in the USA were organized by our
Institute and for many years there were close relations through
visiting professors and postgraduate research fellows. Professor
Andrew Keller presented one of the first reports in the US on his
sensational discovery of chain folding at our Institute and
Professor Max Perutz, newly crowned with the Nobel Prize, visited
Fankuchen's laboratory to tell the fascinating story of the
hemoglobin work in Cambridge, England. C. V. Raman came from
India, I Sakurada and S. Okamura from Japan, V. A. Kargin and
A. Topchiev from the USSR, Hermann Staudinger and Karl Ziegler
from Germany, the Bergmans and Katschalskys from Israel, Stig
Claesson, Bengt Ranby from Sweden and many other distinguished
scientists from other countries.

Many unexpected and far reaching discoveries were made in the
USA and our Institute participated to the best of the ability of
its members in lectures, seminars, symposia and in organizing
monographs and reference books. Enormous progress was made in the
last three decades through the introduction of new methods to
characterize polymers such as lightscattering, dynamic osmometry,
differential thermogravimetric and mechanical analysis, gel
permeation chromatography, and the various magnetic resonance
methods. At the same time, unexpected new approaches for the
preparation of novel types of polymeric materials were opened such
as the "living polymer" concept of Michael Szwarc and the new world
of fluoro plastics and fluoro elastics.

Already in the early fifties, C. S. Marvel had made adventurous
pioneering steps in the synthesis of linear macromolecules which
contain aromatic units as elements of the backbone chain and work
displaying exceptional properties of thermal stability. However,
because of their high softening range it turned out that the conventional polycondensation method could not be applied for their
smooth and successful preparation. A breakthrough was needed in
order to open the way to new horizons of polymeric materials having
properties equal and superior to those of glass and even steel.

This breakthrough was accomplished by P. M. Morgan[13] of the
Textile Pioneering Division of DuPont with his interfacial polymerization technique. Literally hundreds of aromatic polyesters,
polyamides, polyureas and other similar structures were now
suddenly accessible, could be characterized in their structure and
studied for possible applications. Already the first venture in
this direction resulted in a fiber former of considerable interest--
the poly (meta⁻phenylene⁻iso⁻phthalamide) -- known commercially as
Nomex -- which has a softening range above 400°C, and permits to
produce textile fibers of satisfactory strength and of superior
resistance against ignition, flame-spread and heat generation.
However, as often in the area of unrestricted pioneering work, a
close relative of Nomex -- the poly (para-phenylene-diamine-tere-
phthalate) -- now commercially known as Kevlar provided the real
reward for the effort spent. This material, with a softening point
above 500°C, resisted every attempt for dissolution until
Mrs. Stephanie Dwolek found that a sample with adequate molecular
weight can be dissolved and spun from concentrated sulfuric acid
and until Dr. Herbert Blades succeeded to lift the tensile
properties of such fibers to a level almost three times as high as
the best polyamide or polyester fiber. These results, achieved
by a carefully organized team of inventive and motivated experts,
are probably ushering in a new era for the use of fibers, films,
coatings, and all kinds of molded articles because they are equivalent in stiffness, strength and thermal resistance to glass and
metals and are far superior to these classical materials from the
point of view of weight. Already successful applications in bullet
proof vests, motor and sail boats, radial tires and aircraft wings
are a reality and, very probably, soon tests for elevator cables,
ski lifts and suspension bridges will explore their feasibility for
such future uses. However, beyond the already existing and still
forthcoming important practical applications of their "superfibers,"
Morgan and his associates did more than that: they discovered a
new principle -- namely the liquid crystal spinning. During the
preparation and the extrusion of their spinning dopes, they
observed an extreme sensitivity of these systems against shear
caused by a strong tendency for the formation of molecular bundles
even in the fluid state. This pronounced self-orientation of the
individual chains into larger aggregates produces a peculiar condition known as the nematic state which on spinning forms fibers
of an extraordinary degree of molecular order along and across the
fiber axis, which leads to excessive rigidity and strength.

Many areas of polymer research and technology have reached
the state of semisaturation where numerous marginal improvements
are still necessary but where no major evolutions can be expected;
however, there are still disciplines where a great deal of
additional fundamental research is needed to find out what organic
polymers can do in new fields of exploration and endeavor. One of
them is their use in the vast area of hygiene, medicine, and safety,

as auxiliary articles, hard and soft implants, long time delivery systems, artificial organs and blood plasma substitutes. Another attractive and promising field is the use of organic polymers as electro and photoresponsive materials in such systems as electronic conductors, semiconductors, photoconductors, and photochromic, photodielectric, piezoelectric, and pyroelectric devices.

REFERENCES

1. High Polymers, Vol. I, Interscience Publishers (1940), edited by H. Mark and G. S. Whitby.
2. Journal of Polymer Science, Vol. I, Interscience Publishers (1946), edited by H. Mark and P. M. Doty.
3. High Polymers, Vol. II, Interscience Publishers (1940), edited by H. Mark and A. V. Tobolsky.
4. High Polymers, Vol. III, Interscience Publishers (1954).
5. V. Frilette, J. Hanle, and H. Mark, *J.A.C.S.*, 70, 1107 (1948).
6. T. Alfrey, A. Bartovics, and H. Mark, *J.A.C.S.*, 64, 1557 (1942). A. Sookne, H. Rutherford, H. Mark, and M. Harris, *J. Research National Bureau of Standards*, 29, 123 (1942). T. Alfrey, G. Goldfinger, and H. Mark, *J. Applied Phys.*, 14, 700 (1943).
7. S. Tessler, N. Woodberry, and H. Mark, *J. Polymer Sci.*, 1, 437 (1946).
8. H. Mark, *Ing. Eng. Chem.*, 34, 1343 (1942). T. Alfrey and H. Mark, *J. Applied Phys.*, 14, 700 (1943).
9. High Polymers, Vol. VI, Interscience Publishers (1947), Turner Alfrey, Jr.
10. T. Alfrey, E. Merz, and H. Mark, *J. Polymer Sci.*, 1, 37 (1946). H. Mark, *Angew. Chem.*, 61, 313 (1949).
11. High Polymers, Vol. VIII, Interscience Publishers (1952), T. Alfrey, J. J. Bohrer, and H. Mark.
12. H. Mark, *Record Chem. Progress*, 12, 139 (1951). H. Mark, *Research*, 4, 167 (1951). H. Mark, *Textile Research Journal*, p. 294-298 (1952).
13. Polymer Reviews, Vol. 10, Interscience Publishers, P. W. Morgan.

C. S. Marvel

When I began research in organic chemistry, polymers were very poorly understood. Staudinger was just beginning to proclaim them as macromolecules rather than aggregates which were held together in some mysterious manner. There were no techniques such as infra-red spectroscopy, nuclear magnetic resonance and mass spectrometry to use as routine tools for structure determinations. It was then necessary to use chemical methods.

My early research was orthodox organic chemistry and my first entrance into the polymer field was accidental in connection with a reaction between tri-ethyl n-butyl arsonium bromide and lithium ethyl which we ran, hoping to obtain a penta alkyl arsine which had once been tentatively reported by Cahours.[1] We did not obtain a penta alkyl arsenic derivative but a gas was liberated which proved to be very pure ethane. There is no good way to account for ethane to be produced without an equal amount of ethylene being formed. So we postulated that the excess lithium ethyl had caused the ethylene to polymerize. We showed this to be possible by treating a solution of lithium ethyl in a hydrocarbon solvent with ethylene. The ethylene disappeared rather rapidly and a white poly-meric solid separated which was apparently linear polyethylene. At that time, polyethylene was not considered to be very important so no further work on this polymerization was carried out.[2]

My next encounter with polymeric materials came when I was studying the preparation and reaction of a series of halogenated alkyl dialkyl amines.[3]

$$X-(CH_2)_n-N\diagup^{CH_3}_{\diagdown CH_3}$$

These materials underwent a self reaction to give both cyclic and linear polymeric quanternary ammonium salts. The ratio of these two types

$$(CH_2)_n N^+(CH_3)_2 \ X^- \qquad X-\left[(CH_2)_n-\underset{CH_3}{\overset{CH_3}{N}}-(CH_2)_n\right]_y-\underset{CH_3}{\overset{CH_3}{N}} \ \overline{X}$$

varied with the value of n, when n was 4 or 5 cyclization was essentially complete. When n was 3, 7, 8, 9, or 10 the ratio of cyclic compound to linear polymer was about that which Carothers[4] had found to be the case in the self reaction of ω-hydroxy acids.

In 1928, I became a consultant to the Central Chemical Department of the duPont Company. In an early discussion with Dr. Elmer Bolton he called my attention to a British Patent issued in 1914 to Mathew and Elder[5] claiming that sulfur dioxide reacted with simple olefins such as ethylene and propylene to give polymeric products. Bolton asked me if that seemed to be a reasonable reaction and I said it seemed unlikely to me but if he wished I would try it out and see what happened. We were not equipped with pressure equipment to handle ethylene so I tried the reaction with cyclohexene,[6] and found that under peroxide initiation the reaction ran readily to give interesting polymers. About the same time, I found references to work by Solonina[7] describing polymers of allyl derivatives with sulfur dioxide. And, in 1933, both Seyer and King in Canada[8] and Staudinger in Germany[9] reported poly-sulfones from olefins and sulfur dioxide.

The work on sulfur dioxide olefin polymers was continued in my laboratory for about 10 years. We found that the reaction was general for all 1-olefins unless the double bond was conjugated with a carbonyl or a nitrile group. Many 2-olefins also gave poly-mers but if the double bond was buried further in the chain, polymers were usually not formed. Functional groups not conjugated with the double bond did not interfere with the polymerization. 1-Acetylenes also reacted with sulfur dioxide to give polysulfones. All polymers made in our laboratory had a 1-1 ratio of olefin (or acetylene) to sulfur dioxide except vinyl chloride which gave a

polymer containing 2 units of vinyl chloride for each unit of
sulfur dioxide.

When this work was first done, the mechanism of vinyl polymerization had not been established and we were interested in
finding out the exact structure of the polymers formed from an
unsymmetrical olefin. Propylene-polysulfone was selected for study
of the structure.[10] It was found that this polymer was hydrolyzed
readily by alkali to yield a four-carbon two-sulfur derivative and
a polymer of acetaldehyde. The two potential polymer structures
were I and II.

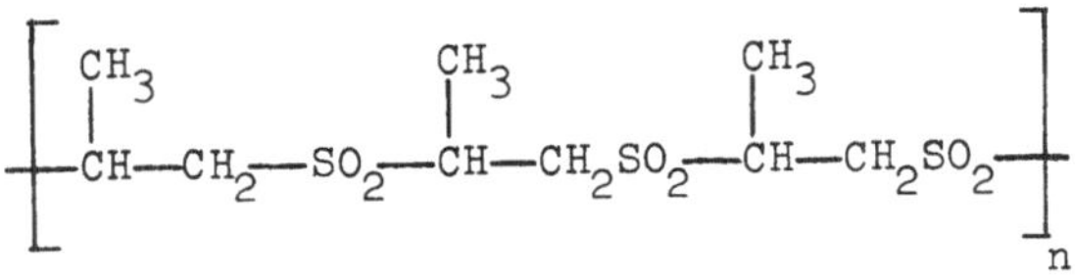

$$\left[-CH-CH_2-SO_2-\overset{\overset{\displaystyle CH_3}{|}}{CH}-CH_2SO_2-\overset{\overset{\displaystyle CH_3}{|}}{CH}-CH_2SO_2- \right]_n$$

I

$$\left[CH_2\overset{\overset{\displaystyle CH_3}{|}}{CH}-SO_2-\overset{\overset{\displaystyle CH_3}{|}}{CH}-CH_2-SO_2-CH_2-\overset{\overset{\displaystyle CH_3}{|}}{CH}-SO_2- \right]_n$$

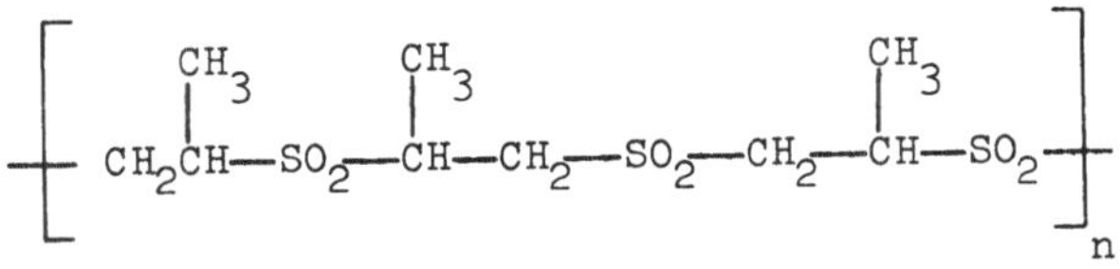

II

Hydrolysis of I might be expected to yield

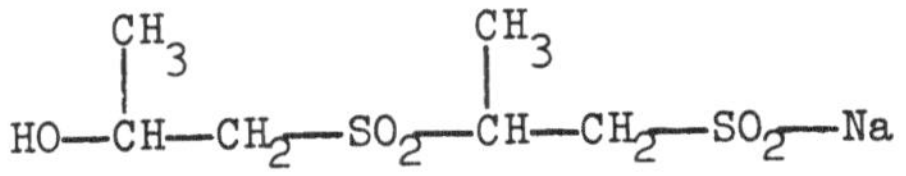

$$HO-\overset{\overset{\displaystyle CH_3}{|}}{CH}-CH_2-SO_2-\overset{\overset{\displaystyle CH_3}{|}}{CH}-CH_2-SO_2-Na$$ and a retrograde aldol reaction

would yield $$CH_3-SO_2-\overset{\overset{\displaystyle CH_3}{|}}{CH}-CH_2-SO_2Na$$ and acetaldehyde.

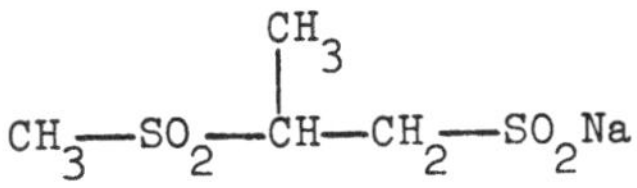

III

Hydrolysis of II might be expected to yield

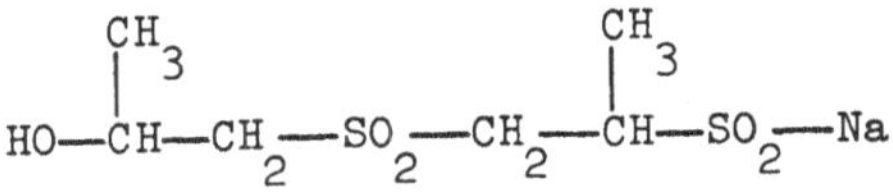

$$HO-\overset{\overset{\displaystyle CH_3}{|}}{CH}-CH_2-SO_2-CH_2-\overset{\overset{\displaystyle CH_3}{|}}{CH}-SO_2-Na$$ and a retrograde aldol reaction

$$CH_3$$
$$|$$
would yield $CH_3SO_2CH_2CH\text{--}SO_2Na$

IV

Two compounds were synthesized which we believed to have structures III and IV and we concluded that the product of polymer hydrolysis had structure IV and therefore the original polymer must have had the head to head, tail to tail structure.

When the mechanism of free radical polymerization was established, this structure seemed improbable. So further work was undertaken;[11] and it was learned that in our synthetic approach to the two model sulfone sulfinic acids rearrangements had occurred and the two synthesized products were actually the reverse in structure to what we had thought. The olefin sulfur dioxide polymers then actually have the orthodox head to tail structures.

The polymers in this series are very beautiful, colorless products which, on melting and molding, give clear "poker chips" much like Lucite (polymethyl methacrylate) except there was always at least one and sometimes more defects due to decomposition of the polymer into gaseous products. No stabilization additives were ever found to make it possible to mold these products without defects and gradually work on the problem was abandoned by us. It has been taken up by several other groups at various times since then but as far as I am aware no one has found a way to stabilize these materials against decomposition at the melting point.

In the early days of polymer chemistry before the mechanism of vinyl polymerization was understood, there was considerable speculation as to structures of vinyl polymers as it was not known whether the units were regularly head to tail, head to head, tail to tail, or random in arrangement. In the late thirties and early forties, we studied the reactions of many vinyl polymers such as polyvinyl chloride, polyvinyl alcohol, polymethyl vinyl ketone and polystyrene to try to settle this point.[12] All proved to be head to tail structures as was later predicted to be correct from the polymerization mechanism theory. Again I emphasize that, at this time, infrared spectroscopy was in its infancy and NMR was unknown so structures needed to be established by chemical means.

In the early forties we tried to work with optically active monomers and optically active initiators to see if we could influence the properties of vinyl polymers which should have active centers in them.[13] We did not succeed in introducing any activity in the polymer chains but did find these monomers were of some help in determining polymerization kinetics for such active monomers.

In 1941 when the supply of East Indian rubber was cut off by the Japanese, the government synthetic rubber program was instituted and I was drafted to help in that program. We organized a research group at Illinois with H. R. Snyder, R. B. Carlin, R. L. Frank, N. Rabjohn, and C. C. Price helping me with the organic problems which arose. F. T. Wall helped with physical chemical problems and H. A. Laitinen with analytical problems. We were one group along with many others from industry and other universities working together on this urgent problem of providing a rubber to replace natural rubber. This entire group cooperated closely and with very little jealousy to slow the advances which were made. So it is very difficult to assign priority for many of the inventions. My group at Illinois studied new comonomers to use with butadiene to see if any were as good as or better than styrene; we studied modifiers to regulate chain length to prevent crosslinking; we studied the structure of the copolymers, especially the ratio of 1, 2, and 1, 4 units of butadiene in the chain; we studied the inhibition effect of polyunsaturated acids in the soap used as emulsifiers; we looked for new units to use in the chain to provide oil resistance without too much sacrifice in the Tg of the polymer; we studied the polymerization of butadiene and its copolymerization with styrene by sodium initiation and also by cationic initiation; we examined new initiation systems; we looked for tackifiers to improve the processability of the synthetic polymer and many other problems.

At the end of the war, I went to Germany with a team of rubber chemists to learn what progress the German chemists had made in synthetic rubber. There we learned about the redox system they had developed for fast polymerization of butadiene and styrene. Their efforts were mainly directed toward a continuous polymerization system. On our return to America, the rubber group here adapted this redox system to low temperature polymerization and the present process for the polymerization of styrene and butadiene in emulsion at $0°$ to $5°$ was developed. This greatly improved the quality of our GRS rubber and made it very satisfactory for use in building passenger car tires.

AN EARLY TYPICAL REDOX RECIPE

Butadiene	75.0 pt.
Styrene	25.0 pt.
Water	200 pt.
Mixed alkane sulfonates	7.5 pt.
Benzoyl peroxide	0.5 pt.
Sorbose	0.6 pt.
Ferric laurate	0.05 pt.
Tetra sodium pyrophosphate ($10H_2O$)	0.5 pt.
Ferrous ammonium sulphate ($6H_2O$)	0.5 pt.

During this period of examining new monomers to use with butadiene we discovered that unsaturated molecules such as benzalacetophenone,

$$C_6H_5CH{=}CHCOC_6H_5, \quad C_6H_5CH{=}CH{-}COCH_3, \quad C_6H_5CH{=}CHCO_2CH_3,$$

benzalacetone, esters of cinamic acid and related monomers with thiophene and pyridine units replacing the phenyl groups would give good copolymers with butadiene. We also found that anthracene and some of its derivatives would also give good copolymers. These types of unsaturated monomers had not previously been studied in vinyl polymerization. Styrene was also found to polymerize with anthracene.

In the late forties and early fifties, we extensively investigated the formation of polymers from dimercaptans and diolefins in emulsion systems with radical initiators. This was then extended to include the preparation of poly thiol esters by the addition of dithiol acids to diolefins.

Also polymercaptals and polymercaptols were obtained from dithiols and aldehydes and ketones. Polydisulfides were prepared by air oxidation of

$$HS-(CH_2)_x-SH + RC\overset{O}{\underset{H}{<}} \longrightarrow H\left[S-(CH_2)_x-S-\overset{\overset{H}{|}}{\underset{\underset{R}{|}}{C}}-S-(CH_2)_x-S-\overset{\overset{H}{|}}{\underset{\underset{R}{|}}{C}}\right]_n$$

dimercaptans. I don't believe any of these polymers ever achieved any practical importance but these and other polymers studied did serve to train a great many students in the techniques of polymer chemistry and that result was probably much more important than any of the polymers that we made.

At the close of the activities in connection with the synthetic rubber program, I became interested in thermally stable polymers which were needed in the airplane industries and related uses. The Materials Laboratory of Wright Patterson Air Force group furnished me financial support for this work. After several starts we found that polyaromatic heterocycles were the best types of thermally stable polymers for use. An early example of this type of product was the PBI made from diphenyl isophthalate and tetraaminodiphenyl which we described first in 1961.[14]

This polymer was of interest to the Materials Laboratory and they had it developed by the Celanese Company to give probably the best heat resistant and fire resistant polymer that has been made. Its expense has prevented it from ever becoming very widely used. But it is oxidatively stable to about 300°C and it will not support combustion although it will burn slowly with direct flaming. It has a tensile strength comparable to a good nylon and it makes a comfortable fabric to wear because it transports moisture.

The discovery of this material opened the way for an immense flood of papers on thermally stable materials which have been appearing in the world since 1961. In this country, Stille at Iowa and Van Deusen and his coworkers at the Materials Laboratory have been busy. In Russia, Korshak, Koton and Berlin have contributed many new products. In Japan, Iwakura and Yoda have been studying these types. Irdis Jones in England and deGaudemaris and Sillion in France have found new and useful products. These are only a few of the many people who have contributed to the developments in this field. We, too, have made many new types of aromatic hetero-cycles. It is fairly simple now to design polymers that are oxidatively stable to about 300°C. Above that temperature most every organic compound begins to oxidize slowly in air. A serious problem is the fabrication of these materials into useful articles since they all are extremely intractable polymers which do not melt easily and are mostly quite insoluble. Moreover, all are expensive and difficult to make. But this has been and is still an extremely active research field in polymer chemistry.

At the present time, my group is working on the problem of preparing thermally stable, processable adhesive resins for use in making composites. The problem that is most difficult is to find a reaction to heat-set a fairly low melting plastic so that it gives a rigid material that is still strong and has some extensibi-lity and strength without the liberation of gaseous byproducts in making the change from a thermoplastic resin to a thermoset. Cross-linking is one way of achieving the transformation but this has the risk of producing a product that is brittle. The formation of new intermolecular cyclic units is a method recently described by Arnold and Hedberg[15] of the Materials Laboratory which raises the Tg of the plastic product to a marked degree and this type of change should not increase brittleness. This is a technique to be further studied in the next year.

As to future programs I do not expect to be undertaking any new fields of polymer chemistry in my remaining active years. I shall probably continue in the thermally stable polymer field.

REFERENCES

1. *Ann.*, <u>122</u>, 337 (1862).
2. *J. Am. Chem. Soc.*, <u>52</u>, 376 (1932).
3. *J. Am. Chem. Soc.*, <u>52</u>, 280 (1930); <u>55</u>, 753 (1933); <u>55</u>, 1977 (1933).
4. *Chem. Rev.*, <u>8</u>, 353 (1931).
5. *C.A.*, 7971 (1915).
6. *J. Am. Chem. Soc.*, <u>56</u>, 7815 (1934).
7. *J. Russ. Phys. Chem. Soc.*, <u>30</u>, 82 (1898).
8. *J. Am. Chem. Soc.*, <u>55</u>, 3140 (1937).
9. Die Hochmolecularen Organische Verbindungen; Julius Springer, Berlin 1932, page 3, footnote 4.
10. *J. Am. Chem. Soc.*, <u>51</u>, 169 (1935).
11. *J. Am. Chem. Soc.*, <u>76</u>, 61 (1954).
12. *J. Am. Chem. Soc.*, <u>60</u>, 280 (1938).
13. *J. Am. Chem. Soc.*, <u>62</u>, 3499 (1940).
14. *J. Poly. Sci.*, <u>50</u>, 511 (1961).
15. U. S. Patent Application 526, 192, Nov. 22, 1974, C.A. <u>83</u>, 180257 (1975).

B. Maxwell

As a mechanical engineer my first interest when I entered the polymer field in 1946 was the application properties of plastics. This then led to studies of polymers in the melt state to better understanding processing and fabrication operations. Hence the interaction between processing and application properties became a natural field of great interest. Although my background was mechanical it became apparent as early as 1948 that the structure-property relationship would become the dominant intellectual discipline that would lead to understanding these materials and would be the guiding theme for future research.

Past Research

In 1946 the principle means of measuring mechanical properties of polymers were the test methods derived from older technologies such as metals and ceramics. They had to be revised for these relatively new materials but little atttention had been payed to the question of whether or not these materials were so different that a new approach should be taken. We asked ourselves what does one really need to know about polymers to apply them properly.

Early studies of the failure process[1,3,6] in glassy polymers, crazing, placed emphasis on the idea that one could separate stress effects from strain effects by taking advantage of the face that even in the glassy state the mechanical response was strongly time dependent. Further studies in the area of variable rate impact testing[2,4,5] and variable rate hardness testing[7] gave clear indication that the time dependence of the response of polymers dom-

143

inated their mechanical behavior and therefore any determination
of their application properties must include viscoelastic considerations.

Clearly some simple method must be devised to characterize the
mechanical response as a function of time (and of course temperature). Creep tests were a possibility but the continued strain
with time would cause changes in structure and thereby make the
structure-property relationship difficult to interpret. Stress
relaxation was also a possibility but at that point there were
instrumentation difficulties in obtaining short time measurements.
What was needed was some simple method that would cover a wide
range of time scale of loading in which the applied strain magnitude was a controlled variable. This line of thought led to the
development of the rotating beam dynamic tester[13].

Many studies were undertaken using this technique[8,12,24,31]
leading to the conclusion that the mechanical loss factor as a
function of frequency and temperature gave the clearest description of how viscoelastic response related to the application
characteristics of polymers. In addition since the strain magnitude could be controlled in the experiment it was possible to determine the limit of linear viscoelastic response and its relationship to internal structure[30].

At the same time that these application properties studies
were going on work was initiated on melt flow behavior. At that
time most melt flow studies were carried out at rather low shear
rates. But polymers are processed at high shear rates under high
pressures. We constructed a monstrosly rugged capillary visco-
meter[18] in which pressures in excess of 20,000 psi could be
achieved and flow studies under conditions similar to actual processing operations were undertaken.

It became apparent immediately that two factors deserved
further consideration. Did the hydrostatic pressure itself affect the viscosity? How to reduce the raw data to meaningful
terms without knowing the real velocity profile in the capillary?

The first question led to the development of apparatus wherein the pressure gradient along the capillary could be applied in
a controlled manner at the same time that specific absolute pressures were applied[19]. The experiments demonstrated that hydrostatic pressure has a strong effect on viscosity and clearly indicated that this effect must be considered in the analysis of
processing operations.

The question of the velocity profile was also attached directly. A tracer particle technique using transparent capillaries

demonstrated that the classical assumption of no slip at the wall
was not valid and in addition that the higher the molecular weight
the more the velocity profile approached plug flow[33],[36]. These results indicate that the design of polymer processing equipment is
really much more complex than indicated by the usual assumptions
of a power law fluid.

The tracer particle technique was later applied to the flow
region at the exit from an extrusion die[40]. The actual velocity
profiles before and after the die exit when combined with equations
of continuity forced the conclusion that density changes were taking
place in this region. This is an important factor in the explanation of die swell.

The earlier work on pressure effects on viscosity when combined with considerations of the molecular mechanisms of melt flow
leads to the conclusion that polymer melts must be highly compress-
ible. Accordingly research was initiated on the bulk compressibility of melts. The most significant result of this work dealt with
pressure induced crystallization[16],[23]. Since pressure could cause
crystallization at temperatures well above the melting point it
would seem that this must be going on in injection molding operations. The second conclusion was that pressure is a homogenous
"nucleating agent" as compared to thermal nucleation by heat transfer. This then gives a method of process induced morphology control and experiments were introduced to see what process induced
structural changes could be achieved by pressure induced crystallization. Results on linear polyethylene indicated a two-fold increase in flexural modulus. This was the beginning of our work on
the relationship between processing, internal structure, as controlled by processing, and application properties. This was followed shortly by the demonstration that pressure induced crystallization could increase the stress cracking resistance of linear
polyethylene by a factor of greater than fifteen[29].

All of these melt studies were based on consideration of the
melt as a liquid, but the early work on application properties
where from the solid mechanics point of view. Elastic effects were
also very important in the melt state. Consideration of the viscoelastic characteristics of polymers in both the liquid and solid
states led to the investigation of the elastic, or solid like, behavior of polymer melts. Studies of the recoverable strain in
polymer melts[28] indicated that if the molecular weight was high
enough the melt exhibited almost a one-hundred percent recovery.
Obviously true understanding of melt processing required consideration of both the viscous and elastic behavior.

The ideal experiment would be one in which the viscous response was separated from the elastic response even though both

were themselves viscoelastic and that this should be done in a homogeneous shear field. This approach led to the orthogonal rheometer[37]. Two forces at 90° to each other in the plane of shear were measured as well as the normal force perpendicular to the plane of shear, hence the name "orthogonal rheometer". Although developed to study structure-property relationships in polymer melts, this orthogonal shear field caused a considerable flurry amongst fluid mechanics and applied mathematics groups.

Many papers were written by others (some are too tensorized for me to read) analyzing this shear field but the eventual consensus was that it measured what we said it did. Meanwhile we went on with our melt studies; the effects of molecular weight and molecular weight distribution[42,43] the effect of branching[46,47], and the stress relaxation of melts after steady shearing in which it was demonstrated that the relaxation spectrum was a function only of the previous shearing frequency[41,44]. The limit of linear viscoelastic behavior in the melt state was studied as a function of molecular structure[52] and the changes of viscous and elastic response as a function of composition in polymer blends demonstrated that there was a maximum in the elastic response[49].

During the period in which these rather academic studies were going on another development was taking place that would have a profound affect on the future course of my research. In 1958 we began to wonder about the utility of the results of the studies on polymer melts. How could we apply the principles being developed? The result was the elastic melt extruder[25,48,54,55]. Here instead of fighting the elasticity of the melt we used it to cause the material to extrude itself. The result was a method of mixing, compounding and extrusion at relatively lower temperatures and shorter residence times that was applicable to all thermoplastics, even those of very high molecular weight and melt elasticity.

This method of mixing and extruding permitted the formulation of polymer blends wherein the individual components were continuous phases. For example a formulation of one percent polyethylene, ninety-nine percent polystyrene showed a continuous phase of polyethylene when the polystyrene was extracted with toluene. Many different blend systems were investigated and this subject continues active in current research.

In the early 1960's we also became interested in the relationship between flow during processing and the kinetics of crystallization. Preliminary studies of crystallization of quiescent melts were undertaken from the point of view of that used by metallurgists, the T-T-T diagram[34,35]. Although most studies of crystallization kinetics are conducted on undisturbed melts the actual crystallization in most processing operations takes place during a shear

experience or shortly thereafter. The early studies indicated that
deformation had an influence on the kinetics and studies directed
to this point were undertaken. It was found that the rate of
crystallization changed by several factors of ten when shear was
applied to the melt as compared to the quiescent melt[38,45,50,56].
The conclusion was that shear acted as a homogeneous nucleating
agent. Here again we have a powerful tool to control morphology
and thereby properties.

Present Research

Studies continue on melt rheology. All of the early work with
the orthogonal rheometer was carried out in the linear viscoelastic
range, relatively small values of applied strain. Real processing
operations involve large magnitudes of strain. In order to bring
our work in line with practical engineering considerations, studies
have been initiated on melts above the linear range. The search is
for a method of describing the viscoelastic response at strains
large enough for the strain itself to have changed the internal
structural response mechanisms. In the orthogonal rheometer shear
field it is possible to vary the magnitude of the strain applied at
constant time scale of loading, i.e., frequency. It has been found
that when large strains, many orders of magnitude above the linear
range, are applied and the strain is then reduced, a linear rela-
tioship exists between the dynamic stress and dynamic strain[57].
It therefore becomes possible to characterize the viscoelastic response above the limit of linear viscoelastic response by a linear
relationship that is itself a function of the maximum strain applied and the frequency or time scale of loading.

Another area of current research deals with combining the earlier
work on production of polymer blends[55] in the elastic melt extruder
and the observation of the maximum in melt elasticity as a function
of composition as found in the orthogonal rheometer studies on
blends[49]. It would be expected that as the melt elasticity increased the forces required for melt orientation would also increase. This has been confirmed in current studies of melt tension
in fiber spinning experiments. The important point is that the
composition of the blend can be used to control the molecular orientation.

It is well known that polymer melts exhibit thixotropic behavior, i.e. reversible structural break down with flow. This
means that an opportunity exists to take advantage of this reduction
in resistance to flow to process polymers with less expenditure of
energy. Current studies of the time dependence of recovery of
structure from previous shearing indicate that rather long times,
up to ten minutes, are available before complete recovery[58]. Ap-

plications of this natural physical phenomena have been demonstrated
by subjecting the melt to sequential shear fields, first one in
which the structure is temporarily broken and then a second shear
field where it is processed into the final product shape.

Future Research

The study of the chemistry of large molecules has dominated the
field. Tremendous advances have been made in the understanding of
the mechanism of the synthesis process in the past thirty years.
On the basis of this fundamental understanding the polymer industry
has grown at an unprecidented rate. Concurrently great advances
have been made in molecular characterization and in the understanding of the morphology of the aggregation of large molecules. The
science of the field has received the best effort of many researchers
and the job has been and is being well done. Now we must turn our
efforts to the application, the engineering of these materials.

Metallurgists have long recognized that processing to produce
finished products must also produce the internal structure that
gives the desired application properties. For example in the drawing of copper wire for electrical conductors as the process continues to produce the final dimensions the grain size, degree of work
hardening and conductivity must all be controlled to optimize the
properties of the final product. In this field the technology has
advanced to the stage where all these factors can be handled at very
high rates of production.

Relatively speaking the polymer field has not yet reached this
stage, to some degree in fibers and films but even here there is
much room for improvement. We have many tools to help us produce
the optimum structure, and optimum properties; pressure induced
crystallization, shear induced orientation and crystallization,
polymer blend composition to control melt rheology as well as the
synthesis processes to produce the desired relationship between internal structure and application properties. This is the area of
future research that deserves our attention.

REFERENCES

1. "Rheological Properties of Polystyrene below 80°C," *Industrial and Engineering Chemistry*, Sept. 1949.
2. "Impact Testing of Plastics: Elimination of the Toss Factor," *A.S.T.M. Bulletin*, October 1949.
3. "Factors Affecting the Crazing of Polystyrene," *Society of Plastics Engineers Journal*, G, No. 9, 1950.

4. "The Effect of Velocity on the Tensile Impact Properties of Methyl Methacrylate," with J. P. Harrington, *A.S.M.E. Transactions*, 74, No. 4, 579, May 1952.

5. "Tensile Impact Properties of Some Plastics," with J. P. Harrington, R. E. Monica, *S.P.E. Journal*, Dec. 1952.

6. "Crazing of Polystyrene," *A.S.T.M. Bulletin*, Dec. 1953.

7. "Hardness Testing of Plastics," *Modern Plastics*, 32, 9, May 1955.

8. "The Dynamic Mechanical Properties of Polymethyl Methacrylate over a Wide Frequency Range," *J. of Poly. Sci.*, 17, 83, May 1955.

9. "The Mechanical Engineering of Dielectrics," *Electrical Engineering*, Vol. 74, Oct. 1955.

10. "Torsional Properties of Plastics," with E. B. Sharp, *Modern Plastics*, 33, 4, Dec. 1955.

11. "A Review of Developments in Plastics Engineering 1954-1955," *Mechanical Engineering*, 78, 6, 533.

12. "An Investigation of the Dynamic Mechanical Properties of Polymethyl Methacrylate," *J. of Poly. Sci.*, 20, 96, 551, June 1956.

13. "An Apparatus for Measuring the Response of Polymeric Materials to a Sinusoidal Oscillating Strain," *A.S.T.M. Bulletin*, 215, July 1956.

14. "Mechanical Properties of Plastics Dielectrics," *Electrical Manufacturing*, p. 146, Sept. 1956.

15. "A Review of Developments in Plastics Engineering 1955-1956," *Mechanical Engineering*, 161, 79, 2, Feb. 1957.

16. "Bulk Compressibility of Polymers at Fabricating Temperatures," with Shiro Matsuoka, *S.P.E. Journal*, Feb. 1957.

17. "Torsional Properties of Plastics," with E. P. Sharp, translated into German by W. Weidmann, *Kunststoffe*, 46, 5, May 1956.

18. "Flow Behavior and Turbulence in Polyethylene," with R. F. Westover, *S.P.E. Journal*, 13, 8, Aug. 1957.

19. "Hydrostatic Pressure Effect on Polymer Melt Viscosity," with Alex Jung, *Modern Plastics*, 35, 3, p. 174, Nov. 1957.

20. "Review of Rubber & Plastics Session, A.S.M.E. 1957," *Rubber World*, Jan. 1958.

21. "Transactions of Society of Rheology," 1957, (Editor).

22. "Flow Behavior & Turbulence in Polyethylene," with R. F. Westover, *Bell Telephone Systems Publication - Monograph 3004.*

23. "Response of Linear High Polymers to Hydrostatic Pressure," with S. Matsuoka, *J. of Polymer Sci.*, 32, 124, 131-159, Oct. 1958.

24. "The Comparison of Time Dependent Mechanical Properties of Plastics," *S.P.E. Journal*, 15, 6, June 1959.

25. "An Elastic Melt Extruder," with A. J. Scalora, *Modern Plastics*, October 1959.

26. "Cold Working of Polyethylene," with Paul H. Rothschild, *Journal of Applied Polymer Science*, Vol. V, No. 16, 511.

27. "Postgraduate Education in Plastics in the United States,"
     Transactions of the Plastics Institute (London), Oct. 1961.
28. "New Light on Melt Elasticity," with Robert A. McCord,
     Modern Plastics, 39, 1, p. 116, Sept. 1961.
29. "You Can Control Brittleness of High Density Polyethylene,"
     with T. S. Brazier, *Modern Plastics*, 39, 2, p. 125, Oct. 1961.
30. "Dynamic Mechanical Spectra and Limit of Linear Viscoelasticity
     of High Polymers," with Claude Guimon, *J. of App. Polymer Sci.*,
     6, 19, 83-93 (1962).
31. "An Evaluation of Polypropylene by Dynamic Mechanical Tests,"
     with James E. Heidner, *Transactions of the Society of Plastics
     Engineers*, 18, 4, (1962).
32. "Efectos Termicos en los Circles de Moldeo de los Termo-
     plasticos," with J. Aleman Vega, *Quimica E Industria*, 8,
     126-131 (1961).
33. "Velocity Profiles for Polyethylene Melt in Tubes," with
     J. C. Galt, *J. of Polymer Sci.*, 62, 174 (1962).
34. "Structure-Property Modifications by Fabrication Operations,"
     with C. G. Gogos, L. L. Blyler, Jr., R. M. Mineo, *Transaction
     of Society of Plastics Engineers*, 4, 3 (1964).
35. "Modifying Polymer Properties Mechanically," <u>Chemical
     Engineering Progress Symposium Series</u>, 60, 49 (1964).
36. "Velocity Profiles for Polyethylene Melts," with J. C. Galt,
     Modern Plastics, 42, 4 (Dec. 1964).
37. "Studies of a Polymer Melt in an Orthogonal Rheometer," with
     R. P. Chartoff, *Transactions of the Society of Rheology*, 9,
     1, 41-52 (1965).
38. "Morphological Foundations of Polymer Processing," *J. of
     Polymer Sci.*, Part C, No. 9, 43-60 (1965).
39. "Measurements of Relaxation Time of a Polyethylene Melt," with
     S. Matsuoka and A. Aloisio, *J. of Polymer Science*, A-2,
     Vol. 4, 113-119 (1966).
40. "Velocity Profiles of the Exit Region of Molten Polyethylene
     Extrudates," with C. G. Gogos, *Polymer Engineering and Sci.*,
     Oct. 1966.
41. "Dynamic Properties of Polymer Melts in an Orthogonal
     Rheometer," *Polymer Science and Engineering*, Vol. 7, No. 3,
     July 1967.
42. "Dynamic Mechanical Properties of Polymer Melts: Effect of
     Molecular Weight on Temperature Dependence of Viscoelastic
     Properties," with R. P. Chartoff, 1968.
43. "Dynamic Mechanical Properties of Polymer Melts: The
     Dependence of Viscoelastic Properties on Molecular Weight
     Distribution," *Polymer Eng. and Science*, 9, No. 3, 159-163,
     May 1969, with R. P. Chartoff.
44. "The Dynamic Behavior and Stress Relaxation of Polymer Melts,"
     Polymer Engineering and Science, 8, 4, Oct. 1968.
45. "Effects of Shear Stress on the Crystallization of Linear
     Polyethylene and Polybutene-1," with T. W. Haas, *Polymer Eng.
     and Science*, 9, No. 4, 225-241, July 1969.

46. "Temperature Dependence of the Properties of Low Density Polyethylene," with D. W. Verser, *Polymer Engineering and Science*, 10, 122 (1970).

47. "Viscoelastic Properties of Branched Polyethylene Melts," with R. P. Chartoff, *J. of Polymer Science*, 8, A-2, 3, 455-465 (1970).

48. "Scaling Up the Elastic Melt Extruder," *S.P.E. Journal*, 26, June 1970.

49. "Dynamic Melt Properties of Polymer Blends," with A. S. Hill, *Polymer Engineering and Science*, 10, 5, 289, Sept. 1970.

50. "Melting of Linear Polyethylene Crystallized from the Melt under Shear Stress," with T. W. Haas, *J. of Applied Polymer Science*, 14, 2407 (1970).

51. "A New Injection Molding System," *Society of Plastics Engineers Journal*, 28, 24-27 (February 1972).

52. "The Limit of Linear Viscoelastic Response in Polymer Melts," with L. H. Gross, *Transactions of the Society of Rheology*, Vol. 16, No. 4 (1972).

53. "Some Observations on Polymer Melt Rheology," Proceedings of Princeton University Conference.

54. "The Application of Melt Elasticity to Polymer Processing," *Polymer Engineering and Science*, Vol. 13, No. 3 (1973).

55. "A New Twist in Mixing Extruders," *Plastics Engineering*, May 1974.

56. "An Experimental Study of the Kinetics of Polymer Crystallization During Shear Flow," with R. R. Lagasse, *Polymer Engineering and Science*, 16, 3, March 1976.

57. Ph.D. Thesis, Gershon Lidor, Princeton University, 1976.

58. Ph.D. Thesis, Lin Choon Lim, Princeton University, 1975.

59. Ph.D. Thesis, Richard Yeh, Princeton University, 1975.

H. Morawetz

My research activities have been largely concentrated in studies of: 1. solid state polymerization, 2. kinetics of reactions in systems containing polymers, and 3. rates of conformational transitions and isomerizations in flexible polymer chain molecules.

SOLID STATE POLYMERIZATION

In 1953, Dr. G. J. Dienes of the Brookhaven National Laboratory asked me to meet with a group at Brookhaven who wanted to explore the possible use of radioactive products of nuclear fission for applications in polymer science. This meeting led to a cooperative program with Dr. R. B. Mesrobian and myself in which ionizing radiation was used for the initiation of graft polymerization and initiation of polymerization reactions in crystalline monomers. After Dr. Mesrobian's departure from our Institute in 1954, this program concentrated on solid state polymerizations which was conceived as a tool for the study of the mobility of molecules in molecular crystals. Initially, there was wide-spread skepticism about the existence of solid state reactions since the polymerization of a substance such as acrylamide, with the relatively low melting point of 84°, might have taken place in a small volume of a liquid phase formed during the exothermic process. Such an interpretation was rendered unlikely by the observation that small concentrations of propionamide in solid solution in the acrylamide crystal reduced drastically the molecular weight of the polymer obtained. Also, in experiments in which the radicals were introduced into the crystals by irradiation at very low temperature before the sample was warmed to temperatures at which polymerization

153

took place, the chain length of polymers obtained was independent
of the radiation dose, indicating that chain termination by bimolecular radical interaction was negligible.[1] A still more convincing proof of the role of the crystal structure was obtained with
salts of polymerizable acids; potassium acrylate was many orders
of magnitude more reactive than the sodium salt, although the
acrylate ion is the reactive species in both cases.[2]

Our main interest was directed to the question whether the
geometric order of the crystalline reagent can be utilized to
impose a preferred direction on the chain propagation. Such
topotactic effects were, in fact, observed in a number of cases.[3]
(a) After irradiation of large single crystals of barium
methacrylate the addition of the second monomer unit to the
$(CH_3)_2CCOO^-$ radical (formed by addition of a hydrogen atom to
the monomer) was followed by ESR spectroscopy. The spectrum was
shown to depend on the orientation of the crystal in the magnetic
field, indicating that the dimer radical was highly oriented with
respect to the crystallographic directions of the parent crystal.
(b) Crystals of p-benzamidostyrene polymerize spontaneously on
heating as much as 80° below their melting point. Studies of
infrared dichroism during the polymerization of single crystals
showed that the orientation of the hydrogen-bonded network of the
amide groups is left virtually undisturbed if the solid state polymerization is carried out below the glass transition temperature
of the polymer. (c) During the polymerization of single crystals
of vinyl stearate, the orientation of the side chain crystallites
of the polymer is identical to the side chain orientation in the
monomer crystal. (d) It was particularly surprising that highly
oriented polymer chains can be produced by solid state polycondensation. A single crystal of ε-aminocaproic acid is converted
30° below its melting point to an assembly of Nylon-6 crystallites
which exhibit 3-dimensional orientation relative to the crystallographic directions of the monomer crystal.[4] This reaction is not
observed if the monomer vapor is continuously removed under high
vacuum and this indicates that the polymerization involves interaction of the monomer vapor with the monomer crystal surface.

KINETICS OF REACTIONS IN SYSTEMS CONTAINING POLYMERS

(a) <u>Activation of polymer side chains by neighboring groups
attached to the polymer</u>. The solvolysis of certain active ester
groups was found to be accelerated by several orders of magnitude
by suitably spaced ionized carboxyl groups.[5] With methacrylic acid
copolymers, this effect depends on the relative steric configuration
of the backbone carbons carrying the ester and carboxylate groups
and this effect provided the first experimental method for the
estimation of the "microtacticity" of the chain. (In fact, the

term was first used in the context of this study.)[5b]

(b) Interaction of non-neighboring groups attached to flexible chain molecules. Reaction kinetics were studied in solutions of chain molecules carrying reactive and catalytic chain substituents at long spacings along the chain backbone. The reaction rate is here a measure of the statistical probability of cyclic chain conformations in which the two interacting groups are in juxtaposition, i.e., a measure of the chain flexibility. Of particular interest was the influence of the excluded volume effect on the probability of cyclization.[6]

(c) Polyelectrolyte catalysis of ionic reactions. If two interacting species are attracted into a polymer domain, the probability of mutual collision of the reagents will be increased and the polymer will then act as a catalyst for the reaction. Using this principle for the reaction of two cationic species in solutions of polymeric acids, acceleration rates of up to five orders of magnitude were observed. The reaction is proportional to the polyion concentration as long as only a fraction of the reagents are bound to the polymer domain. However, when the polymer concentration exceeds the value at which virtually all the reagents are polymer-bound, the reaction rate decreases with further polyion addition, since the reagents have to be shared by an increasing number of polymer domains, so that their concentration decreases in any one of them.[7] A particularly striking effect was observed in a study of the quenching of UO_2^{++} fluorescence by Fe^{++} in solutions of poly (vinylsulfonic acid). At low concentrations of Fe^{++}, the polyanion increased the quenching efficiency, since both the fluorescing and the quenching ions were drawn into its domain. However, as the Fe^{++} concentration was increased, the UO_2^{++} from the polyion domain to regions in which the concentration of the quenching species is low.[8]

(d) Reactions between groups attached to two different flexible chain molecules. In a solution containing two similar polymers, carrying small concentrations of reactive and catalytic chain substituents, respectively, it was expected that the resistance of the polymer coils to mutual interpenetration in good solvent media would reduce the reaction rate below the value observed in systems containing low molecular weight analogs of the interacting species. No such effect was observed[9] and although no satisfactory explanation of this result has been formulated, it must be considered an important piece of evidence which should be considered in theories of polymer solutions.

(e) Enzymatic attack on side chains of flexible polymers. Some evidence in the literature suggests that pharmacologically active substances are inactive when attached to polymers but become available to the organism as they are detached from the

macromolecule by enzymatic cleavage. We have investigated the
attack of chymotrypsin on polymer side chains as a function of the
spacing of the susceptible bond from the chain backbone and the
nature of this backbone.[10]

   (f) Reactivity of groups attached to a crosslinked polymer
network. The Merrifield synthesis of polypeptides with a uniquely
defined sequence of amino acid residues has had a dramatic impact
on the feasibility of preparation of physiologically active
materials such as polypeptide hormones. Yet, little is known about
the dependence of the reactivity of a group attached to a polymer
gel on the nature of the polymeric support. Work in our laboratory
on a model system has shown that the reactivity of side chains is
profoundly influenced by the nature of the polymer backbone and
that variations in the topology of a polymer network can lead to
large changes in reactivity even if the swelling behavior of the
gel in the reaction medium is kept constant.[11]

RATES OF CONFORMATIONAL TRANSITION AND ISOMERIZATION
IN FLEXIBLE POLYMER CHAINS

   It has been suggested that conformational transitions in the
backbone of polymer chains involve two correlated hindered rotations
in a so-called crankshaft-like motion. If this is the case, the
free energy of activation for such hindered rotations in the backbone of a polymer should be significantly higher than in low
molecular weight analogs and such rotations should, therefore, be
slower in the polymers. We have used an NMR method to study
hindered rotation in polyamides[12] and ultraviolet spectroscopy to
follow rates of photochemical and thermal isomerization of
azobenzene residues in the backbone of polymer chains, but found
no difference in dilute solution between the behavior of the
polymers and their analogs.[13,14] Studies were also carried out in
concentrated systems.[14,15] In the glassy state, the thermal
isomerization of polymer side chains exhibits a characteristic
deviation from first-order kinetics, reflecting the microheter-
ogeneity of the glassy state.[15] The efficiency of the photochemical
isomerization of azobenzene residues in the polymer backbone
decreases rapidly with decreasing plasticizer content to extremely
low values.[14]

   Intramolecular excimer fluorescence involves a conformational
transition furing the lifetime of an excited chromophore and it
can, therefore, be used to study conformational transitions on the
timescale of 10^{-8} sec. We have used this technique to study the
behavior of copolymers[16] and to define the dependence of the rate
of conformational transitions on the viscosity of the medium.[17]

FUTURE PLANS

The following investigations are either currently active or projected for the near future:

(a) Intramolecular excimer formation is studied as a function of the bulk of the groups involved in the conformational transition and the viscosity of the medium. It is also planned to use intra-molecular excimer fluorescence to characterize the microviscosity in plasticized polymer compositions.

(b) A reflectance fluorescence method will be developed to follow the kinetics of the reaction of groups attached to cross-linked polymer networks. We hope to define conditions under which a resin, of potential use in the Merrifield synthesis, should be prepared so as to minimize the dispersion of the reactivities of groups attached to the polymer network.

(c) The stereoselectivity of an asymmetric reaction yields information about the difference in non-bonded atom interaction energies in the two diastereomeric transition states. We have recently used this approach to study the effect of various additives to water on the magnitude of these non-bonded interaction energies. This program should be expanded to a study of the stereoselectivity of adsorption on and catalysis by chiral resins.

(d) A recent crystallographic study of side-chain crystalliza-tion[18] left many questions unanswered. It is planned to expand this study to a wider range of polymers and to supplement x-ray crystallography by electron microscopy to gain more information about the organization of the side-chain crystallites.

REFERENCES

1. T. A. Fadner and H. Morawetz, *J. Polymer Sci.*, 45, 475 (1960).
2. H. Morawetz and I. D. Rubin, *J. Polymer Sci.*, 57, 669 (1962).
3. H. Morawetz, *Pure Appl. Chem.*, 12, 201 (1966).
4. E. M. Macchi, N. Morosoff, and H. Morawetz, *J. Polymer Sci.*, Pt. A-1, 6, 2033 (1968).
5. (a) P. E. Zimmering, E. W. Westhead, Jr., and H. Morawetz, *Biochem. Biophys. Acta*, 25, 376 (1957); (b) H. Morawetz and E. Gaetjens, *J. Polymer Sci.*, 32, 526 (1958); (c) E. Gaetjens and H. Morawetz, *J. Am. Chem. Soc.*, 83, 1738 (1961).
6. N. Goodman and H. Morawetz, *J. Polymer Sci.*, Pt. C, 31, 177 (1970); *J. Polymer Sci.*, Pt. A-2, 9, 1669 (1971).
7. (a) H. Morawetz and B. Vogel, *J. Am. Chem. Soc.*, 91, 563 (1969); (b) H. Morawetz and G. Gordimer, *J. Am. Chem. Soc.*, 92, 7532 (1970).

8. I. A. Taha and H. Morawetz, *J. Am. Chem. Soc.*, <u>93</u>, 829 (1971).
9. (a) H. Morawetz and W. R. Song, *J. Am. Chem. Soc.*, <u>88</u>, 5714
    (1966);   (b) J.-R. Cho and H. Morawetz, *Macromolecules*, <u>6</u>,
    628 (1973).
10. T.-Y. Fu and H. Morawetz, *J. Biol. Chem.*, (1976), in press.
11. H. Morawetz, in <u>Peptides: Chemistry, Structure, and Biology</u>,
    R. Walter and J. Meienhofer, eds., Ann Arbor Science Publ.,
    1975, P. 385.
12. Y. Miron, B. R. McGarvey and H. Morawetz, *Macromolecules*, <u>2</u>,
    154 (1969).
13. D. Tabak and H. Morawetz, *Macromolecules*, <u>3</u>, 403 (1970).
14. D. T.-L. Chen and H. Morawetz, *Macromolecules*, (1976) in press.
15. C. S. Paik and H. Morawetz, *Macromolecules*, <u>5</u>, 171 (1972).
16. Y.-C. Wang and H. Morawetz, *Macromol. Chem.*, *Suppl. 1*, 283
    (1975).
17. Y.-C. Wang and H. Morawetz, *J. Am. Chem. Soc.*, (1976) in press.
18. H. W. S. Hsieh, B. Post, and H. Morawetz, *J. Polymer Sci.*,
    Pt. A-2, (1976), in press.

M. Morton

My major research activities of the past and present cover six areas of polymer science, as follows:

1. Emulsion polymerization
2. Urethane polymerization
3. Polymerization of cyclic siloxanes
4. Anionic polymerization
5. Anionic synthesis
6. Reinforcement of Elastomers by Model Polymeric Fillers

The above, which are arranged in roughly chronological order, represent the main areas of activity, and do not include minor research projects which have "spun off" from some of these. The attached bibliography identifies the various publications resulting from this work.

I. EMULSION POLYMERIZATION

1. <u>Thiol Transfer Kinetics</u>

During the late 1940's, as an offshoot of the U. S. Government Synthetic Rubber Program, some of the first data were obtained during my stay at McGill University on the chain transfer kinetics of thiols (specifically the tertiary alkane-thiols) in emulsion polymerization of butadiene-styrene rubber. These demonstrated for the first time the relation between the rate constants of a series of such "transfer agents" (C_8 to C_{16}) and the molecular weight

159

distribution of the polymers at various conversions. It therefore
laid the groundwork for effective control of molecular weights in
in these systems.

2. Crosslinking in Polymerization of Dienes

As an outcome of the above interest in molecular weight
control of butadiene-styrene copolymers, we turned our attention,
at Akron, during the late 1940's and early 1950's, to the problem
of "gelation" which is encountered in these polymerizations.
Following the original systematization of "three-dimensional" polymerizations by Flory and Stockmayer, a kinetic analysis of the
special case of crosslinking in diene polymerization was developed
by Flory[1]. This was then used, in conjunction with the above-
mentioned knowledge available about the kinetics of chain transfer
in emulsion polymerization to evaluate, for the first time, the
rates of the crosslinking reaction in the polymerization of butadiene, and later, of isoprene and 2,3-dimethylbutadiene. (This
approach was not successful in the case of chloroprene.) This
information had a visible impact, as one might expect, on the understanding of the profound effect of temperature on the formation of
insoluble networks ("gel") in synthetic rubber production, at a time
when the industry was converting rapidly to low temperature polymerization ("cold rubber"). It still stands today as the only
comprehensive data on these important reactions.

3. Absolute Propagation Rates in Free Radical

Diene Polymerization. Simultaneously with the resolution of
the question of the crosslinking reaction in emulsion polymerization
of dienes, it became possible to evaluate the rate constants of the
polymerization reaction itself. This was, of course, primarily due
to the definitive analysis of the kinetics of emulsion polymerization
by Ewart and Smith[2]. By a careful study of various "recipes" used
in the emulsion copolymerization of butadiene-styrene, it was found
that certain low-temperature systems obeyed Case II of the Ewart-
Smith treatment, i.e.,

$$R_p = k_p M(N)/2$$

where: R_p = rate of polymerization

k_p = propagation rate constant

M = monomer concentration

(N) = number of latex particles per unit volume

Hence, it was thus possible to evaluate k_p, the absolute propagation rate constant from a measurement of R_p, M, and (N). Incidental to this, the determination of M and its changes during growth of the latex particles led to the elucidation of the thermodynamics of swelling equilibria in such systems (see Ref. 11 in the attached bibliography).

The determination of the absolute propagation rate constants for the free radical polymerization of butadiene, isoprene and, later, 2,3-dimethylbutadiene, was especially important since these monomers are not amenable to the classical method for such measurements, i.e., photopolymerization by ultraviolet radiation, which leads to the complicating side-reaction of crosslinking of the polydienes. Thus, for the first time, information became available about these rate constants, and this, in turn, made it possible to calculate the _absolute_ values of the other reactions in free radical systems, e.g., termination, transfer and crosslinking. For example, values for the propagation and crosslinking rate constants are shown in Table I for the three dienes. Termination rate constants are not shown, since only that for butadiene was determined (unpublished results), and found to be of the order of 10^9 M^{-1} sec^{-1} at 60°C., i.e., approximately two orders of magnitude higher than that of styrene. This inordinately high value for the termination rate helps to explain the unusual tendency of the 1,3 dienes to give very slow rates and very low molecular weights in homogeneous free radical systems, a fact which was noted by the early investigators in synthetic rubber research.

TABLE I

ABSOLUTE RATE CONSTANTS FOR BUTADIENE, ISOPRENE AND DMBD

Monomer	Reaction	A [a]	E	k_{60} [b]
Butadiene	Propagation	1.2×10^8	9.3	100
Isoprene	Propagation	1.2×10^8	9.8	50
DMBD	Propagation	8.9×10^7	9	120
Butadiene	Crosslinking	1.9×10^9	16.8	0.020
Isoprene	Crosslinking	8.5×10^6	14.8	0.0017
DMBD	Crosslinking	3.0×10^5	14.4	0.0011

a) $M^{-1}S^{-1}$ b) KG. CAL.

II.   URETHANE POLYMERIZATION

During the early 1950's, the "hottest" item in polymer synthesis was probably the urethane polymer system, as developed by Otto Bayer[3] and his associates, since this led to a whole new concept of "chain extension" of oligomers to high polymers by the ingenious use of the high reactivity of isocyanate groups towards labile hydrogen atoms, e.g., as in hydroxyl and amine groups. The Vulcollan types of "castable" rubbers, for example, was the direct outcome. It soon became known that these systems also exhibited a wide variety of "side reactions," due to the high reactivity of the isocyanate groups, and involving such moieties as water, urethane groups, biuret groups, etc., which are present either as impurities or as reaction products. Hence, we felt it important to establish more exactly the relative rates of these reactions, which can have a profound influence on such characteristics as molecular weight, branching, etc., and which would strongly influence the properties of the final product.

The rates of some of these competitive reactions of urethane chemistry, as determined from model compound studies, are shown in Table II. It can be seen that such knowledge is of vital importance in understanding the relative rates of chain extension, branching and crosslinking in the reaction of diisocyanates with both simple and polymeric diols, diamines, etc.

TABLE II[4]

RELATIVE REACTIVITIES OF PHENYL ISOCYANATE

Reactant	Type of Polymer Reaction	Relative Rate
n-Butyl Carbanilate (a urethane)	Branching	1
n-Butyranilide (an amide)	Branching	14
Carbanilide (a urea)	Branching	74
n-Butyric Acid	Chain extension	78
Water	Chain extension	295
n-Butanol	Chain extension	1375

III. POLYMERIZATION OF CYCLIC SILOXANES

During the early and mid-1950's, much interest was also aroused by studies on the base-catalyzed polymerization of cyclic siloxanes, such as octamethylcyclotetrasiloxane. The pioneering work of Grubb and Osthoff[5] on the polymerization of the undiluted monomer hèlped greatly in elucidating the anionic nature of these systems. We thought it important to explore these reactions further and to study such factors as solvent and temperature effects on both the rates and molecular weights. In this way we were able to show the profound solvent effects which one can expect in such anionic systems, and to demonstrate, by molecular weight measurements, the non-terminating character of the chain growth reaction.

Since this was also the time when the non-terminating ("living") nature of certain carbanionic polymerizations was discovered, it is not surprising that this knowledge was applied to the siloxane systems. Thus potassium naphthalene was found in our laboratory to be an effective initiator for polymerization of the cyclic tetrasiloxane, and, at the same time, block copolymers of styrene or isoprene and the siloxane were successfully synthesized.

IV. ANIONIC POLYMERIZATION

1. Molecular Weights of "Living" Polymers

Two events, both of which occurred in the mid-1950's, had a profound effect on the work in our laboratories. The first was the discovery[6], in the Firestone research laboratories, that lithium or organolithium initiators could polymerize isoprene to a very high cis-1,4 configuration, thus almost duplicating the structure of natural rubber. The second was the announcement[7] that styrene could be polymerized by sodium naphthalene under conditions in which no chain termination or transfer occurred, thus leading to almost monodisperse and predictable molecular weights. These discoveries opened up a whole new field of interest in anionic polymerization, at the same time as the breakthrough in stereospecific polymerization by the Ziegler-Natta catalysts.

Turning our attention first to the polymerization of styrene by sodium naphthalene, we showed, in collaboration with the group at the National Bureau of Standards, the type of narrow distribution of molecular weights which could be obtained by this method. We then showed that such "living" polymerizations were not unique with sodium naphthalene, which is restricted to ether solvents, but can be generalized to other homogeneous anionic systems, such as the polymerization of styrene, butadiene, isoprene, etc., by organolithium, both in ether and hydrocarbon media.

2. Kinetics of Organolithium Polymerization

During the late 1950's and early 1960's, in the course of examining the unique relationship between the stoichiometry and the molecular weights obtained in the homogeneous anionic polymerization of styrene and the dienes, our attention was naturally drawn to the mechanism and kinetics of these reactions. In the case of the organolithium systems, we found a profound influence of the solvent on the effect of concentration of initiator (i.e., of growing chain ends) on polymerization rate. Thus, in ethers, the kinetic order, with reference to initiator concentration, ranged between 1/2 and unity, while, in hydrocarbon media, it seemed at first to be strictly 1/2, but was later found to be even lower, depending on the type of monomer and solvent. For example, styrene polymerizations, either in benzene or cyclohexane, always appear to be half-order with respect to organolithium initiator, while those of butadiene or isoprene can be as low as 1/4 or even 1/6 order, depending on solvent.

The rate dependency of anionic polymerizations on initiator concentration, in ethers or other polar media, has been adequately explained as due to the simultaneous propagation of ion-pairs and free anions in equilibrium with each other. However, no such mechanism can be invoked in the case of the non-solvating hydrocarbon solvents. We had already pointed out, in 1960, that in the latter case, one can observe a strong association of chain ends, and our measurements of the viscosity of the polymerizing solutions indicated that these active chains are associated in pairs. It was, therefore, attractive to propose a mechanism in which only the unassociated species was capable of propagation, so that the order of the reaction would be dependent on the order of association, as illustrated by the following equations:

$$(RLi)_n \;\underset{\longleftarrow}{\overset{K_e}{\longrightarrow}}\; nRLi$$

$$R_p = k_p K_e^{1/n} [(RLi)_n]^{1/n}$$

where RLi is a growing chain:

$$R_p \text{ is the rate of polymerization}$$

and this was at first thought to be the case. However, later results showed that although the association number, n, was invariably 2 in all cases measured, i.e., styrene, butadiene and isoprene, the kinetic order was 1/2 only in the case of styrene, and considerably lower for the dienes. Hence this attractive hypothesis had to be abandoned, and an explanation of the very low kinetic orders of the dienes must lie elsewhere.

The complexity of these organolithium polymerizations in hydrocarbon media was amply illustrated by our further studies, which showed that the initiator (alkyllithium) participates in a cross-association with the active chain ends, thus leading to a variety of associated species, presumably having different reactivities toward the monomer.

More promising kinetic data were obtained when the "living" character of the growing chain ends was utilized in measuring directly the absolute values of the "crossover" reaction in the copolymerization of styrene and butadiene. This was conveniently done by simply observing the appearance or disappearance of the peak at 436 nm in the ultraviolet region, which corresponds to polystyryl lithium. In this way the k_{12} and k_{21} values were obtained for this copolymerization system, from which the r_1 and r_2 values were constructed.

3. Studies of the Nature of the Propagating Chain End in Organolithium Polymerization of Dienes

The complexities in the kinetics of organolithium polymerization, especially in hydrocarbon media, as described above, spurred more interest in elucidating the actual structure of the carbon-lithium bond, which comprises the growing chain end during the polymerization reaction. With this objective in mind, we turned to high resolution proton magnetic resonance as a useful tool. These studies were started during the late 1960's and are still continuing today.

The PMR spectra obtained showed the effect of the lithium on the chemical shift of the protons on the terminal monomer unit in a series of polydienes, including butadiene, isoprene and 2,3-dimethylbutadiene, and the effect of solvents on these spectra. The most striking result of this work was to show the effect of solvating solvents, e.g., ethers, in delocalizing the electrons of the carbanionic chain end, from a largely covalent carbon-lithium bond to a π-allylic type. This correlated very well with the activity of the γ-carbon in generating side-vinyl structures in the polymer chain (1,2 units of polybutadiene and 3,4 units of poly-isoprene).

In addition to this demonstration of the effect of solvents on the carbon-lithium bond, this work also showed the effect of substitution on the carbon atom involved. Thus, similar NMR studies of the 1,3-pentadienes, and especially of 2,4-hexadiene showed that a secondary carbanion assumed a π-allylic structure even in hydrocarbons, presumably because of its lower stability (higher energy) as compared to the primary carbanion.

4. Chain Structure of Polydienes

One of the unexplained phenomena in organolithium polymerization is the marked difference between isoprene and butadiene, in
that the former goes to very high cis-1,4 configuration in hydrocarbon media, while the latter does not. It is also known that
the chain structure is affected by the lithium concentration, but
this has never been accurately determined. Very recently, we
carried out studies in our laboratories to pinpoint the effect of
these reaction variables on the polydiene chain structure. These
results showed that even different hydrocarbon media exert an
effect, although not a dramatic one, on the chain structure. Thus
higher cis-1,4 structures are obtained for polyisoprene prepared
in hexane than in benzene, while still higher cis-1,4 contents are
obtained with undiluted monomer. Superposed on this effect is the
influence of the organolithium concentration, the highest cis-1,4
contents being obtained with the lowest organolithium concentration.
Thus, for example, only 70% cis-1,4 content was obtained in benzene
with 10^{-3} M initiator while as high as 94% cis-1,4 resulted from a
polymerization of undiluted monomer with 10^{-5} M initiator.

Similar results were obtained with butadiene, except that the
cis-1,4 contents were not as high, the maximum attainable with
undiluted monomer and 10^{-5} M initiator being about 85%. These
were the first results to relate the chain structure to such subtle
factors in the reaction, and are an indication of the marked
sensitivity of the carbon-lithium bond to the polarity of the
medium.

V. ANIONIC SYNTHESIS

As is well known, homogeneous anionic polymerization is
especially well suited to the synthesis of various and unique
polymers, because of the non-terminating character of the polymerization. The two main avenues that have been explored with
organolithium systems in our laboratories are: 1) block copolymerization, and 2) synthesis of polymers having functional end
groups.

1. Block Copolymers

During the mid-1960's our attention was drawn to the
remarkable properties of the ABA type of "triblock" copolymers,
where the A block is polystyrene and the B block is a polydiene
(butadiene or isoprene). These yield products of the "thermoplastic elastomer" type, since the glassy polystyrene end blocks
aggregate into regular and well-formed domains which act as
thermoplastic network junctions for the elastic polydiene chains.

The first results obtained in our laboratories were aimed at elucidating the structure-property relations which govern the behavior of these polymers. Because we were able to prepare block copolymers of precise, predictable and well-characterized molecular weights, we were able to relate such parameters as composition and block size to the mechanical properties, with special reference to tensile strength. Our most important conclusion was probably that the strength of these elastomers depends mainly on the degree of phase separation between the glassy domains and the rubbery matrix.

As a continuation of this work, various analogs of these triblock copolymers were synthesized, such as α-methylstyrene-b-isoprene-b-α-methylstyrene, α-methylstyrene-b-(propylene sulfide)-b-α-methylstyrene and α-methylstyrene-b-dimethylsiloxane-b-α-methylstyrene. All of these showed similar morphology and structure-property relations as the styrene-diene triblocks, as might have been expected. It was noteworthy, however, that when the polystyrene end blocks were replaced by poly-α-methylstyrene, there was a noticeable increase in modulus and tensile strength, at any given temperature. This was presumably due to the enhanced ability of the poly-α-methylstyrene domains to withstand greater stresses and higher temperatures.

2. Polymerization of Cyclic Sulfides

As indicated above, cyclic sulfides, such as propylene sulfide, were used as monomers in the synthesis of block copolymers. Other cyclic sulfides, such as ethylene sulfide, trimethylene sulfide (thiacyclobutane), and 2=methylthiacyclobutane were also polymerized by organolithium initiators to yield either rubbery or crystalline block copolymers. In the course of this work, some interesting new light was thrown on the mechanism of the reaction between organolithium and cyclic sulfides, as illustrated below.

For 3-membered rings

$$RLi + \underset{\diagdown S \diagup}{CH_2\text{-}CHCH_3} \longrightarrow CH_2{=}CHCH_3 + RSLi$$

$$RSLi + \underset{\diagdown S \diagup}{CH_2\text{-}CHCH_3} \longrightarrow \ldots \longrightarrow RS\text{-}[CH_2\text{-}\overset{\overset{CH_3}{|}}{CH}\text{-}S]_xLi$$

For 4-membered rings

$$RLi + \begin{array}{l} \quad\;\; CH_3 \\ \quad\;\; | \\ CH_2\text{-}CH \\ | \qquad\;\; | \\ CH_2\text{-}S \end{array} \longrightarrow \cdots \longrightarrow R\text{-}[S\text{-}\overset{CH_2}{CH}\text{-}CH_2\text{-}CH_2]_xLi$$

The two noteworthy, and unexpected, phenomena are, of course, the β-elimination reaction involved in the initiation step of propylene sulfide, and the carbanionic nature of the ring-opening reaction of the thiacyclobutane ring. Hence it is possible, for example, to initiate the polymerization of styrene with a block of polythiabutyllithium, since the latter has a carbanionic chain end, rather than a thiolate, as in the case of poly(propylene sulfide).

3. Synthesis of Uniform Polyisoprene Networks

It has long been a dream of rubber chemists to prepare uniform elastic networks, directly, by interlinking the ends of short chains, rather than by random crosslinking of long chains, as is done at present. This, of course, requires the synthesis of disfunctional chains of prescribed length and narrow molecular weight distribution, and this should be possible by the methods of anionic polymerization.

This goal has just now been accomplished in our laboratories. A series of α,ω-dihydroxpolyisoprenes, having very narrow molecular weight distributions, has been prepared, with molecular weights in the range of 3,000-12,000. These were then end-linked into a net-work by means of a tri-isocyanate, with an efficiency of better than 95%. The sequence of reactions is diagrammed below.

Although this study is still in progress, results to-date al-ready indicate that these uniform networks are superior in strength

$$CH_3-CH=CH-CH=CH-CH_3 + Li \xrightarrow{(C_2H_5)_3N} Li[\overset{CH_3}{\underset{|}{CH}}-CH=CH-\overset{CH_3}{\underset{|}{CH}}]_{2-3}Li$$

$$\downarrow C_5H_8$$

$$Li[CH_2-\overset{CH_3}{\underset{|}{C}}=CH-CH_2]_x-[\overset{CH_3}{\underset{|}{CH}}-CH=CH-CH]_{2-3}-[CH_2-CH=\overset{CH_3}{\underset{|}{C}}-CH_2]_x Li$$

where x - 50-200.

$$\downarrow (CH_2)_2O$$

$$HO-[CH_2-\overset{CH_3}{\underset{|}{C}}=CH-CH_2]_x-[\overset{CH_3}{\underset{|}{CH}}-CH=CH-CH]_{2-3}-[CH_2-CH=\overset{CH_3}{\underset{|}{C}}-C-CH_2]_x OH$$

$$\downarrow CH(p-C_6H_4NCO)_3$$

Network

to, and exhibit more ideal elastic behavior than, conventional crosslinked networks, which have a distribution of network chain lengths.

VI. MODEL POLYMERIC FILLER STUDIES

One of the most intriguing phenomena in polymer science is the remarkable effect which fine, particulate fillers, such as carbon black or silica, have in increasing the strength of noncrystallizing elastomers. Although it is generally accepted that the particle size of such fillers must be very small (<500A), it has been difficult to correlate the effect of these fillers with their characteristics, since the latter are quite ill-defined. Furthermore, it is also known that various chemical interactions between the fillers and the rubber can occur during vulcanization, because of reactive chemical groups at their surface.

A research program was therefore started in our laboratories during the mid-1960's involving the use of well-characterized polymeric fillers as reinforcing agents in rubber vulcanizates. The approach used was to prepare a filler (e.g., polystyrene) of desired particle size by emulsion polymerization, to blend it in desired proportion with an SBR latex containing the necessary vulcanizing ingredients, to coagulate the blended, compounded latex and to vulcanize the rubber at temperatures well below the softening point of the polymeric filler.

This experimental approach had the following advantages: a) the fillers are well-defined, relatively uniform in size, and spherical in shape, b) the fillers are well dispersed as individual particles, rather than as aggregates, and 3) the presence of chemical bonds between filler and rubber can be detected by extracting the vulcanizate with solvents.

As a result of these investigations the following features pertaining to the mechanism of filler reinforcement of elastomers have been elucidated:

1) Fillers increase the strength of elastic networks by increasing the viscous component of these viscoelastic materials (hysteresis), and, for the same volume fraction, smaller particles thus have a greater effect. This is deduced from the fact that all of these filled vulcanizates lie on the same "failure envelope" but at different locations.

2) The presence of infrequent chemical bonds between filler and rubber has no noticeable effect on strength.

3) The stiffness, or modulus, of the filler has a direct effect on strength, a more rigid filler being more effective.

4) The extent of strength reinforcement is directly dependent on the "adhesion" between filler surface and the elastomer matrix, as shown by density measurements of the strained vulcanizates. For any given type of filler, the adhesion is greatest with the smaller particles, but it can also be affected by the nature of the filler surface.

FUTURE PLANS

Academic research does not lend itself readily to long-range planning, since one never knows what new development will suddenly capture one's imagination. As for immediate plans, I think that we might, during the next few years, be able to accomplish the following objectives related to the programs described above:

1) Elucidate more fully the changes in the structure of the carbon-lithium bond, in hydrocarbon media, especially as it is affected by its concentration and to relate it to changes in the chain structure and kinetics in diene polymerization.

2) Accumulate more complete data on the effect of network chain size distribution on the mechanical properties of elastomers.

3) Study the effect of particle shape on the filler reinforcement of rubber vulcanizates, using the model polymeric fillers described above.

DESCRIPTIVE CITATIONS

1. P. J. Flory, *J. Am. Chem. Soc.*, 69, 2893 (1947).
2. W. V. Smith and R. H. Ewart, *J. Chem. Phys.*, 16, 529 (1948).
3. O. Bayer, *Rubber Chem. Technol.*, 48, G73 (1975).
4. M. Deisz, *Ph.D. Dissertation, University of Akron*, 1959.
5. W. T. Grubb and R. C. Osthoff, *J. Am. Chem. Soc.*, 77, 1405 (1955).
6. F. W. Stavely and co-workers, *Ind. Eng. Chem.*, 48, 778 (1956).
7. R. Milkovich, M. Levy and M. Szwarc, *J. Am. Chem. Soc.*, 78, 2656 (1956).

REFERENCES

1. W. G. Baker, *Trans. Can. Inst. Mining Met.*, 44, 515 (1941).
2. R. V. V. Nicholls, *Can. J. Research*, 25B, 159 (1947).
3. R. V. V. Nicholls, *Can. J. Research*, 26B, 581 (1948).

4. D. B. MacLean and R. V. V. Nicholls, *Ind. Eng. Chem.*, 41, 1545 (1949).

5. P. P. Salatiello, *J. Polymer Sci.*, 6, 225 (1951).

6. P. P. Salatiello and H. Landfield, *J. Polymer Sci.*, 8, 113 (1952).

7. P. P. Salatiello and H. Landfield, *J. Polymer Sci.*, 8, 215 (1952).

8. P. P. Salatiello and H. Landfield, *J. Polymer Sci.*, 8, 279 (1952).

9. P. P. Salatiello and H. Landfield, *Ind. Eng. Chem.*, 44, 739 (1952).

10. H. Landfield, *J. Amer. Chem. Soc.*, 74, 3523 (1952).

11. S. Kaizerman and M. W. Altier, *J. Colloid Sci.*, 9, 300 (1954).

12. "Branching in Dienes," *Ann. N.Y. Acad. Sci.*, 57, 432 (1953).

13. J. A. Cala and I. Piirma, *J. Polymer Sci.*, 15, 167 (1955).

14. "Molecular Weights of Emulsion Polydienes," *Ind. Eng. Chem.*, 47, 333 (1955).

15. J. A. Cala and M. W. Altier, *J. Polymer Sci.*; 19, 547 (1956).

16. I. Piirma, *J. Polymer Sci.*, 19, 563 (1956).

17. J. A. Cala and I. Piirma, *J. Am. Chem. Soc.*, 78, 5394 (1956).

18. I. Piirma, *J. Am. Chem. Soc.*, 80, 5596 (1958).

19. I. Piirma, *J. Polymer Sci.*, Pt. A-1, 3043 (1963).

20. A. Rembaum and E. E. Bostick, *J. Polymer Sci.*, 32, 530 (1958).

21. P. Cayre and J. P. Berry, *Proceedings of International Rubber Conference, Washington, D.C.*, November, 1959.

22. E. E. Bostick and R. Livigni, *Rubber & Plastics Age*, 42, 397 (1961).

23. J. R. Purdon, *J. Polymer Sci.*, 57, 453 (1962).

24. T. E. Helminiak, S. D. Gadkary and F. Bueche, *J. Polymer Sci.*, 57, 471 (1962).

25. I. Piirma, R. J. Stein, J. F. Meier, *Proceedings of Fourth Rubber Technology Conference*, McLaren Press, London, May, 1962.

26. F. R. Ells, *J. Polymer Sci.*, 61, 25 (1962).

27. R. Milkovich, D. B. McIntyre, L. J. Bradley, *J. Polymer Sci.*, 1A, 443 (1963).

28. A. Rembaum and J. L. Hall, *J. Polymer Sci.*, 1A, 461 (1963).

29. E. E. Bostick and R. G. Clarke, *J. Polymer Sci.*, Pt. A-1, 475, (1963).

30. E. E. Bostick, R. A. Livigni and L. J. Fetters, *J. Polymer Sci.*, A-1, 1735 (1963).

31. L. J. Fetters and E. E. Bostick, *J. Polymer Sci.*, C1, 311 (1963).

32. W. E. Gibbs, *J. Polymer Sci.*, 1A, 2679 (1963).

33. M. P. Wagner, *Rubber Age*, 93, 746 (1963).

34. E. E. Bostick, *J. Polymer Sci.*, 2A, 523 (1964).

35. M. A. Deisz and E. E. Bostick, *J. Polymer Sci.*, A2, 513 (1964).

36. "Big Molecules," Part 1, *Chemistry*, 37 (1), 13 (1964).

37. "Big Molecules," Part 2, *Chemistry*, 37 (3), 6 (1964).

38. L. J. Fetters, *J. Polymer Sci.*, 2A, 3311 (1964).

39. "Real Synthetic Rubber," *Int. Sci. Tech.*, August (1964).

40. "Anionic Copolymerizations," *Copolymerization*, High Polymer Series, G. E. Ham, Ed., Vol. 18, Chapter 7, Interscience, New York, 1964. p. 421.

41. "Mechanism of Stereospecific Polymerization of Butadiene in Aqueous Media," *Rubber & Plastics Age*, 46, 404 (1965).

42. "The Nature of Organometallic Polymerization - Review," *Advan. Chem. Ser.*, 52, 1 (1966).

43. A. A. Rembaum and E. E. Bostick, *J. Appl. Polymer Sci.*, 8, 2707 (1964).

44. *Chemical Reactions of Polymers*, E. M. Fettes, Ed., Volume XIX, Chapter X, High Polymer Series, Interscience, N.Y., 1964, p.811.

45. "Polymers of the Future," *Rubber Age*, 99, 84 (1967).

46. "The Impact of Polymer Science on Synthetic Rubber," *Rubber and Plastics Age*, Dec. 1967.

47. L. J. Fetters, *Chapter in Macromolecular Reviews*, John Wiley and Sons, Inc., 2, 71, 1967.

48. R. L. Denecour, *Proc. Inter. Rubber Conf.*, 1967, MacLaren and Sons, London, 1968, p. 175.

49. *Conference on Advances in Polymer Science and Materials*, Princeton University, Nov. 1968, p. 78.

50. J. C. Healy, *Applied Polymer Symposia*, 7, 155 (1968).

51. B. Das, *J. Polymer Sci.*, C(2), 1 (1969).

52. J. E. McGrath and Peter C. Juliano, *J. Polymer Sci.*, Part C, No. 26, 99 (1969).

53. *Vinyl Polymerization*, G. E. Ham, Ed., Part II, Chapter V, Marcel Dekker, N.Y., 1969.

54. L. J. Fetters, F. C. Schwab, C. R. Strauss, and R. F. Kammereck, *Rubber and Technical Press*, London, 1969, p. 70.

55. L. J. Fetters, *Macromolecules*, 2, 453 (1969).

56. Lewis J. Fetters, R. A. Pett, and J. F. Meier, *Macromolecules*, 3, 327 (1970).

57. Robert A. Pett and L. J. Fetters, *Macromolecules*, 3, 333 (1970).

58. R. F. Kammereck, *J. Am. Chem. Soc.*, 92, 3217 (1970).

59. *Chapter in Block Polymers*, S. L. Aggarwal, Ed., Plenum Press, 1970.

60. "Mechanism of Reinforcement of Elastomers by Polymeric Fillers," *Advan. Chem. Ser.*, 99, ACS, Washington, D.C. (1971).

61. R. D. Sanderson and Ryozo Sakata, *J. Polymer Sci.*, B(9), 61 (1971).

62. Rudolf F. Kammereck and L. J. Fetters, *Macromolecules*, 4, 11 (1971).

63. R. F. Kammereck and L. J. Fetters, *B. Polymer J.*, 3, 120 (1971).

64. *Encyclopedia of Polymer Science and Technology*, John Wiley & Sons, Inc., Vol. 15, p. 508 (1971).

65. *Journal of the IRI*, 6, 148 (1972).

66. J. L. Trout and T.-C. Cheng, *Proceedings of the Int. Rubber Conf.* 1972, Institution of the Rubber Industry, London, p. G6-1.

67. *MTP International Review of Science*, Butterworths, London, 1972, Chap. 1, Vol. 8.
68. *Rubber Chem. & Technol.*, 45(6), G. 78 (1972).
69. L. A. Falvo and L. J. Fetters, *J. Polymer Sci.*, B(10), 561 (1972).
70. L. J. Fetters, *Macromolecular Synthesis*, 4, 77 (1972).
71. *Proceedings of the 4th Buhl International Conference on Materials*, Pittsburgh, PA, Nov., 1971, p. 328.
72. R. D. Sanderson and R. Sakata, *Macromolecules*, 6, 181, (1973).
73. R. D. Sanderson, R. Sakata and L. A. Falvo, *Macromolecules*, 6, 186 (1973).
74. L. A. Falvo, *Macromolecules*, 6, 190 (1973).
75. E. R. Santee and Virgil D. Mochel, *Polymer and Letters*, 11, 45 (1973).
76. E. R. Santee, Jr., and R. Chang, *Polymer Letters*, 11, 449 (1973).
77. M. Morton, E. R. Santee and L. O. Malotky, *Rubber Chem. & Technol.*, 46, 1156 (1973).
78. M. Morton, *J. Chem. Ed.*, 50, 731 (November, 1973).
79. M. Morton, *J. Chem. Ed.*, 50, 740 (November, 1973).
80. M. Morton and S. L. Mikesell, *J. Macromol. Sci.-Chem.*, A7(7), 1391 (1973).
81. M. Morton, R. J. Murphy and T. C. Cheng, *Polymer Preprints*, 15, 436 (1974).
82. L. O. Malotky and M. Morton, *Polymer Preprints*, 15, 714 (1974).
83. M. Morton, *Rubber World*, 169, 14 (1974).
84. L. J. Fetters, *Macromolecules*, 7, 552 (1974).
85. L.-K. Bi and L. J. Fetters, *Polymer Preprints*, 15, 157 (1974).
86. *Polymer Preprints*, 15, 175 (1974).
87. *Chemistry*, 47, 11 (1974).
88. Advances in Synthetic Rubber, Proceedings of the International Symposium on Macromolecules, Elsevier Publishing Co., N.Y., 1975, p. 287.
89. Irja Piirma, V. R. Kamath and Maurice Morton, *J. Polymer Sci.*, 13, 9 (1975).
90. Maurice Morton and Lewis J. Fetters, *Rubber Chem. Technol.*, 48, 3 (1975).
91. M. Morton, R. J. Murphy and T. C. Cheng, Advances in Chemistry Series, No. 142, American Chemical Society, Washington, 1975, p. 409.
92. J. K. Carver and R. W. Tess, Eds., Applied Polymer Science, American Chemical Society, Washington, 1975, p. 161.
93. Maurice Morton, Y. Kesten and L. J. Fetters, *Applied Polymer Symposium*, 26, 113 (1975).
94. M. Morton, L. J. Fetters, J. Inomata, D. C. Rubio, and R. N. Young, *Synthesis and Characterization of α,ω-Dihydroxy-polyisoprene*, 49, 2, 1976.

BOOKS

1. <u>Introduction to Rubber Technology</u>, Reinhold Publishers, New York, 1959.
2. <u>Rubber Technology</u>, Van Nostrand Reinhold Publishers, New York, 1973.

PATENTS

1. Sulfur-Containing Elastomeric Organosiloxane Polymers (with M. A. Deisz), U. S. Patent 2,960,492 (1960).
2. Organosiloxane Polymers Containing Polydiene Blocks (with A. Rembaum), U. S. Patent 3,051,684 (1962).
3. Preparation of Organodilithium Initiator and Polymerizations Utilizing Same (with L. J. Fetters), U. S. Patent 3,663,634 (1972).
4. Rubbery Polymers (with Lewis J. Fetters and Rudolf F. Kammereck), U. S. Patent 3,804,922 (1974).

C. G. Overberger

INTRODUCTION

Over a period of 32 years, our group has published over 338 papers, primarily in the field of organic polymer chemistry. Of the first twelve papers published, the first ten were with Professor C. S. Marvel, who was my thesis advisor. In addition, I spent two years as a postdoctoral fellow in his laboratory after completing the Ph.D. degree.

It is interesting that the first paper published was the synthesis of optically active styrene derivatives and the use of this monomer to check the kinetics of free radical catalyzed vinyl polymerization. Most of this earlier work was directly related to research related to rubber, since we were at war and since there was an extensive synthetic effort to synthesize new rubbery materials. It was during this period that I was introduced to oxidation reduction reactions to initiate vinyl polymerization.

I then spent a year at Massachusetts Institute of Technology where, with Professor A. C. Cope, I repeated the controversial Willstätter synthesis of cyclooctatetraene.

PAST RESEARCH

1. Side Chain Crystallization

From 1947-50, we were interested in the question of how the side chain of an amorphous polymer affected the second-order transition temperature. The general concept of side chain

crystallization occurring if the side chain of the polymer contained long alkyl side chains was resolved. This problem was
undertaken partially in collaboration with Professor Turner Alfrey
as the earlier concepts of graft copolymerization were emerging.
This research clearly demonstrated that in structurally regular
amorphous vinyl polymers, side chain crystallization occurred when
the side chain consisted of straight chain alkyl hydrocarbons of
more than 12 carbon atoms.

2. Catalyst Research

Synthesis and Decomposition of Azo Compounds--Free Radical
Initiators. Our laboratory was one of the first groups to be
heavily involved in the use of azonitriles as free radical
initiators. In particular, we illustrated the fact that virtually
no induced decomposition of these azonitriles occurred upon decomposition in solution. We provided detailed studies of the

effect of structure of the azonitriles on the rate of decomposition.
In particular, we emphasized the stereochemical function on the
rate of decomposition such as front strain and back strain. With
cyclic azonitriles, we were the first to demonstrate the effect of
internal strain in a radical reaction. This research led us to do
a detailed study of the synthesis and

decomposition of cyclic azo compounds. Our initial impetus in
this area was to attempt to experimentally verify whether a free
radical chain could grow from a single source catalyst via two
radical chains simultaneously. We conclusively proved that this
was not the case. I will refer to this topic again under Present
and Future Research. Since diazines are structurally isomeric

Table I

Decomposition Rates of Azo Nitriles in Toluene[60]

$$RR'(CN)C-N=N-C(CN)R'R$$

R	R'	k, sec^{-1} x 10^4 (cor. to 80.0°)
CH_3[a]	$CH_2-C(CH_3)_3$[b]	158 ± 13
CH_3[a]	$CH_2C(CH_3)_3$[c]	136 ± 10
CH_3	$\underline{n}-C_5H_{11}$	1.63 ± 0.10
$i-C_4H_9$	$i-C_4H_9$	49.5 ± 2.5
$i-C_3H_7$	$\underline{i}-C_3H_7$	1.25 ± 0.04
$i-C_3H_7$	C_2H_5	0.95 ± 0.05

[a] Stereoisomers.

[b] $E_a = 27 ± 1.5$ kcal/mole; $\Delta S^\dagger = 7 ± 4$ e.u.

[c] $E_a = 29 ± 1.5$ kcal/mole; $\Delta S^\dagger = 12 ± 4$ e.u.

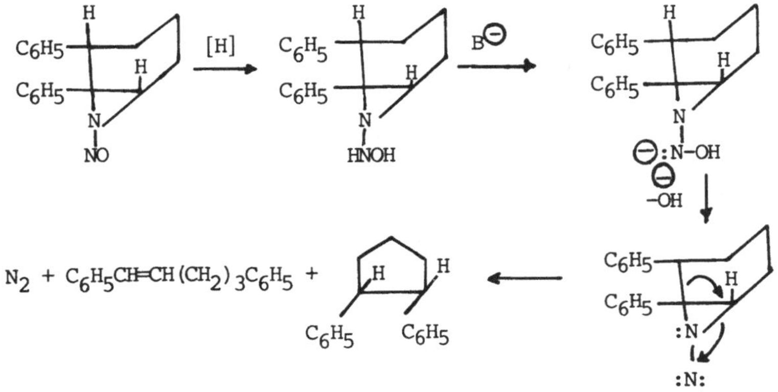

Eq. 3

with cyclic diazo compounds, we also began a systematic study of elimination reactions of diazines. We discovered a new reaction, namely the elimination reactions of N-nitroso compounds when reduced with sodium hydrosulfate. This whole area proved to be very fruitful and we set a firm background and basis for much of the work which is now being carried out in this field.

3. Cationic Catalyzed Polymerization

Our group entered this field at a very early state of the art. Our contributions fall into three main categories:

Table II

Monomer Reactivity Ratios for Alkyl-Substituted Styrenes (V_2)

and p-Chlorostyrene (V_1)

M_2	V_1	V_2
α-Me	0.35 ± 0.05	15.0 ± 1.5
p-Me	0.22 ± 0.05	4.5 ± 0.7
p-C_2H_5	0.29 ± 0.04	4.1 ± 0.5
styrene	0.34 ± 0.05	2.0 ± 0.2
trans-β-Me	0.74 ± 0.06	0.32 ± 0.04
cis-β-Me	1.0 ± 0.1	0.32 ± 0.02
β-C_2H_5	0.88 ± 0.30	0

Eq. 4.1

Eq. 4.2

Table III

Molecular Termination Constants and Relative Reactivities of Monoalkylbenzenes

| Substituent | $k_r/k_p \times 10^3$ | | | Relative Reactivity | | |
	CCl_4 diluent	C_6H_{12} diluent	CCl_4 diluent	C_6H_{12} diluent	Propylation	Bromination
CH_3-	8.8	8.7	100	100	100	100
C_2H_5-	8.0	7.8	90	90	86	76
$(CH_3)_2CH-$	--	6.0	--	69	80	44
$(CH_3)_3C-$	4.5	5.0	50	57	67	23

Table IV

Relative Reactivities of the Methylbenzenes

Reagent	Basicity	Bromination	Iodination	Mercuration
Toluene	0.01	0.33	0.14	0.61
p-Xylene	1	1	1	1
o-Xylene	3	2.3	1.3	2.0
m-Xylene	18	7.1	23	4.2
Mesitylene	1400	625	1800	26
Durene	60	65	23	3.5
Pentamethylbenzene	4300	2580	740	24
Hexamethylbenzene	445000	- -	- -	- -

Table V

Reactivity Ratios r_1 and r_2 for the Copolymerization of

Isobutylene (M_1) and p-Chlorostyrene (M_2) in Various Solvents

Solvents	Dielectric Constant of Solvent	Initiator	r_1	r_2
Benzene	2.28	$SnCl_4$	12.2 ± 1.0	2.8 ± 0.03
Nitrobenzene	36.0	$SnCl_4$	8.6 ± 0.20	1.25 ± 0.1
n-Hexane	1.82	$AlBr_3$	1.1	1.04
Benzene	2.28	$AlBr_3$	1.14 ± 0.10	0.99 ± 0.10
Ethylene dichloride	10	$AlBr_3$	2.80	0.89
Nitrobenzene	36	$AlBr_3$	14.9 ± 0.2	0.53 ± 0.04
Nitromethene	38	$AlBr_3$	22.2 ± 0.5	0.73 ± 0.05
Mixture of benzene-nitrobenzene (50% vol.)	- -	$AlBr_3$	18.0	0.75

a) Copolymerization studies in cationic catalyzed systems. We were able to demonstrate the mechanism of the carbocationic polymerization using substituted styrenes by copolymerization and showing that the reactivities of these monomers obeyed the classical Hammit free energy equation. In addition, we demonstrated steric factors involved in cationic polymerization through the copolymerization method. We demonstrated that β-methyl styrene would copolymerize whereas β-ethyl styrene would not.

b) Our most extensive work in the cationic polymerization area was the elucidation of termination and transfer reactions and the development of what we called molecular termination, namely the reaction of a polymeric carbocation with aromatic entities (an electrophilic substitution). We were able to map clearly the reactivities of polymeric carbocations in electrophilic substitution and compare the reactivities and base strengths of many aromatics with other electrophilic substitutions such as nitration, bromination, mercuration, etc. Of particular note was the conclusion that major steric factors become important with a polymeric cation and alkyl substituted benzenes which are not present when a proton is the reference acid. The question of chain transfer in cationic catalyzed polymerization was clearly elucidated by our contributions.

c) One of the most significant results which arose from our work on cationic catalyzed copolymerization was the effect of solvent on monomer reactivity. We were the first to demonstrate and explain the effect of solvation of both the monomer and solvent on the relative reactivity ratios and explain this phenomena in a systematic manner.

4. Polymers Containing Mercaptan Groups

A large number of novel synthetic procedures were developed for the synthesis of polymers containing mercaptan side chains. Just to name a few, the polymerization of divinyl thiocarbonate, the polymerization of S,S'-divinyl mercaptans, the polymerization of S-, O-, and N-vinyl derivatives of carbonic acid, the polymerization of S-vinyl-O-t-butyl thiolcarbonates, the polymerization of thioureas and isothiouronium salts. We attempted and were only partially successful in relating the structure on the rate of mercaptan oxidation. Many of these polymers containing mercapto side chains were screened for antiradiation agents and some proved to be very promising. Professor Helmut Ringsdorf, who was a collaborator in our group at that time, carried on some of this work at the Universities of Marburg and Mainz, and his contributions have resulted in very definitive antiviral agents specifically active in the intestines.

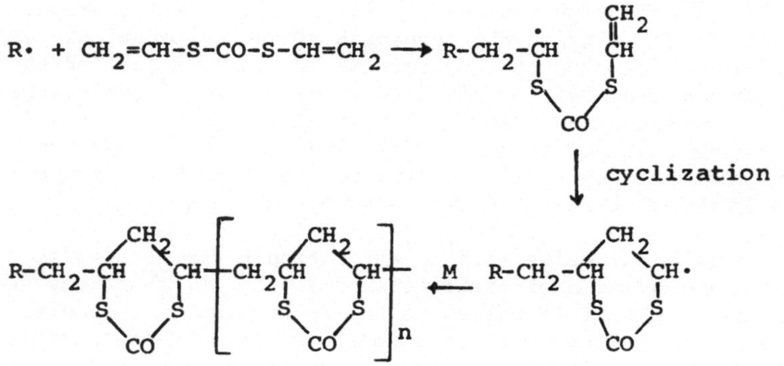

$$R\cdot + CH_2=CH-S-CO-S-CH=CH_2 \longrightarrow R-CH_2-\overset{\cdot}{CH}\quad\overset{CH_2}{\underset{S\quad\quad S}{CH}}$$

$$\text{cyclization} \qquad \text{Eq. 5}$$

$$R-CH_2-CH \overset{CH_2}{\underset{S\quad S}{CH}} CH\!-\!\!\left[CH_2\,CH \overset{CH_2}{\underset{S\quad S}{CH}} CH\right]_n \quad\overset{M}{\longleftarrow}\quad R-CH_2-CH \overset{CH_2}{\underset{S\quad S}{CH}} CH\cdot$$

$$(HOCH_2-CH_2-S)_2CH-R \quad\overset{\text{KOH}}{\underset{200-250°}{\rule{2cm}{0.4pt}}}$$

$$\underset{KO-C(CH_3)_3}{\overset{\text{\underline{tert}-butanol}}{\rule{2cm}{0.4pt}}} \longrightarrow CH_2=CH \overset{H_2C}{\underset{S\quad S}{CH}} \underset{\underset{R}{CH}}{}$$

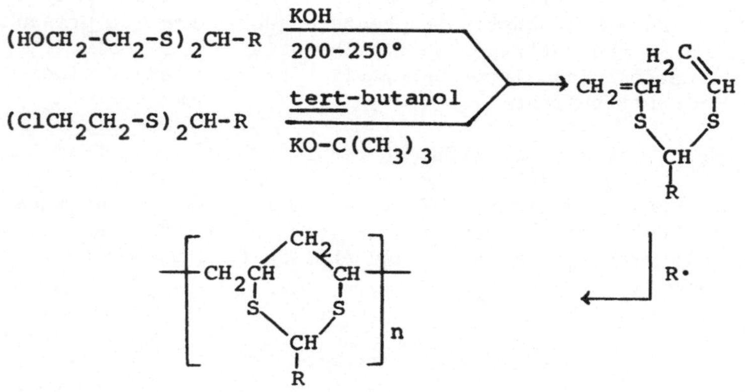

$$\text{Eq. 6}$$

$$\left[CH_2CH \overset{CH_2}{\underset{S\quad S}{CH}} CH\right]_n \quad\overset{R\cdot}{\longleftarrow}$$

Eq. 7

$$CH_2=CH-\!\!\bigcirc\!\!-NH-\underset{\underset{S}{\parallel}}{C}-NR_1R_2$$

$$CH_2=CH-\!\!\bigcirc\!\!-NH-\underset{\underset{S-CH_3}{\overset{\oplus}{C}}}{C}-NR_1R_2 \quad I^{\ominus}$$

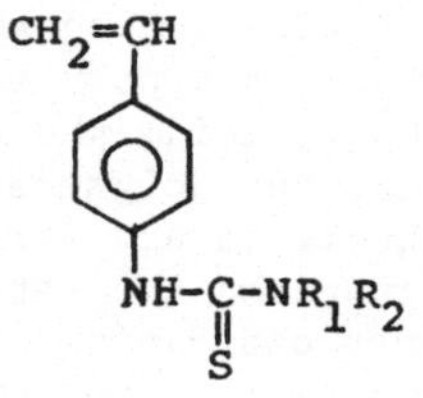

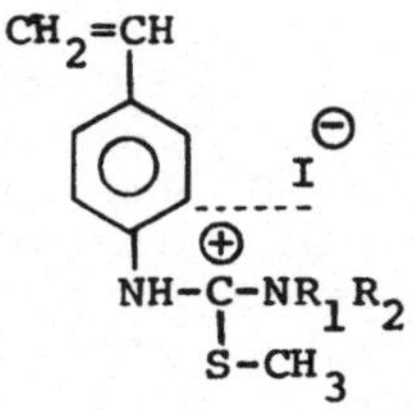

Eq. 8

a) $R_1 = R_2 = H$

b) $R_1 = H,\ R_2 = CH_3$

c) $R_1 = R_2 = CH_3$

Table VI

Polymeric S-, N- and O-Vinyl Derivatives of Carbonic Acid $\{CH_2-CH\}_n$ with R

	R	Softening Point, °C	$[\eta]^{29.5°}_{benzene}$	Soluble in
1	$-S-CO-S-CH(CH_3)_2$	60-80	1.02	benzene,THF
2	$-S-CO-O-C((CH_3)_3$	105-120	0.33	benzene,CHCl$_3$
3	$-N-CO-S-C_2H_5$ / CH$_3$	135-175	1.09	benzene,THF
4	$-O-CO-N$ (ring with CH$_2$-CH$_2$-O-CH$_2$-CH$_2$)	115-135	0.14	benzene,THF

Table VII

Relative Oxidizabilities of Thiols in DMF

Thiol	Observed Oxidation rate $\mu m/mol/min/SH \times 10^5$	$k \times 10^{-3} \ell/mol/sec$	Relative Rate
$CH_3-CH(CH_2)_3CH-CH_3$, SH	0.328	0.0547 ± 0.002	1.0
$HOCH_2-CH_2-SH$	1.70	0.233 ± 0.010	5.2
$CH_3-CH(CH_2)_2-CH-CH_3$, SH SH	3.90	0.650 ± 0.030	11.9
$CH_3-CH-CH_2-CH-CH_3$, SH SH	9.70	1.62 ± 0.10	28.7
$CH_3-CH-CH-CH_3$, SH SH	20.1	3.35 ± 0.20	61.2
$(CH_2-CH-CH_2-CH-CH_2)_{n/2}$, SH SH	91.4	15.2 ± 0.60	280

5.   Anionic Cationic Polymerization

Our most significant work in this area was again in the copolymerization area. We again demonstrated the effect of solvent
and monomer solvation as an important factor in monomer reactivity.
We provided a reasonable explanation for the behavior of styrene
and methyl methacrylate in a copolymerization reaction catalyzed
by lithium metal.

6.   Transition Metal Catalysis

Our particular concern in the area of transition metal
catalysis, particularly heterogeneous catalysis, was again focused
on copolymerization. We demonstrated that steric factors in substituted styrenes played an important role in the reactivity of a
monomer in a transition metal catalyst system. By fractionation
techniques we were able to demonstrate that many copolymers
prepared by a transition metal catalyst were really mixtures of
copolymers with different copolymer composition and composition
distributions. Detailed kinetic studies of the copolymerization
of deuterio substituted styrenes were carried out in a homogeneous
system which contributed to our understanding of the mechanisms of
homogeneous transition metal catalysis.

CURRENT AND FUTURE

1.   Cyclic Azo Compounds

Our work on the preparation and decomposition of cyclic azo
compounds in more recent years is now being gradually phased out.
We were the first to demonstrate the fact that one could have both
a cis and trans azo linkage in an 8- and 9-membered ring.

cis   trans

Eq. 9

Table VIII

Fractionation of a Styrene-4-Methyl-1-Pentene Copolymer

Fraction No.	Fraction wt., g	Fraction wt. %	$[n]_{CCl_4}$ 100 cc/g	St content of fraction, wt. %
1	0.092	4.1	insoluble in CCl_4	7 (5) *
2	0.350	15.4	insoluble in CCl_4	11 (5) *
3	0.248	10.9	sol. in boiling CCl_4	9
4	0.487	21.5	3.1	26
5	0.387	17	0.50	31
6	0.483	21.3	0.07	83
7	0.222	9.8	0.06	95
Total	2.268 (93% of the initial amount)	100.0		

* IR analysis.

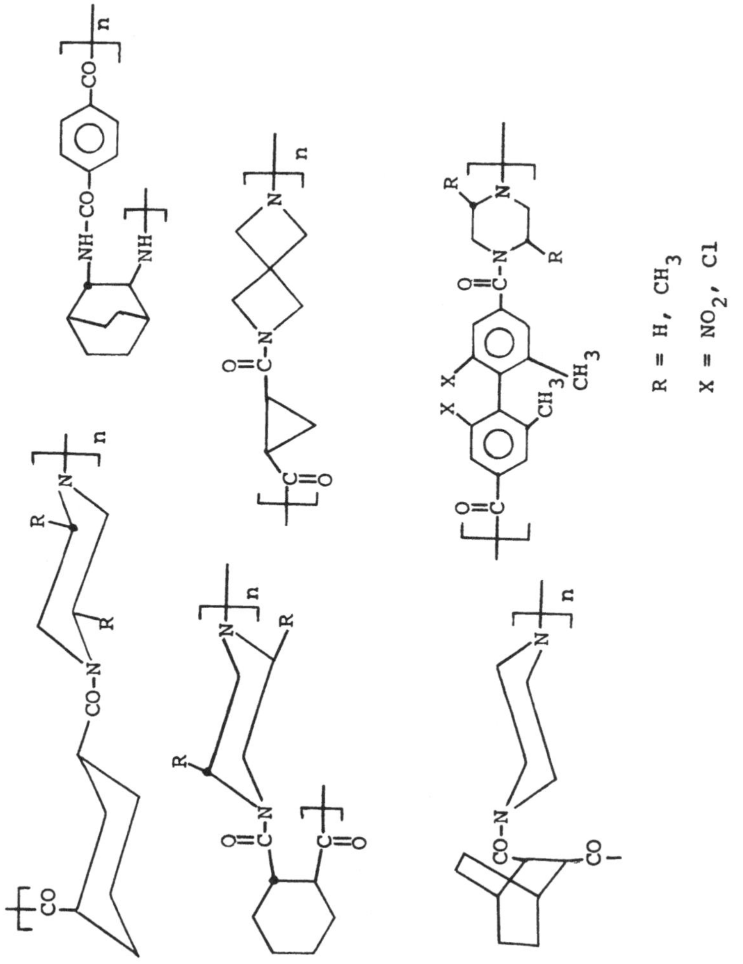

R = H, CH$_3$

X = NO$_2$, Cl

2.   Optically Active Polymers and Their
Conformation in Solution

Our initial goal in the synthesis of asymmetric polymers,
largely polyamides and polyesters, was to determine the effect of
assymmetry on the solution and solid state properties of these
polymers. We were very much aware of the increasing effort on the
study of conformation of polyproline in solution and we believed
that we could also determine the structure and conformation of
other asymmetric polyamides in solution. In a series of over 40
papers we have shown that with the use of high resolution nuclear
magnetic resonance spectrometry, dipolar measurements, simi-
empirical energy calculations, intrinsic viscosity, and circular
dichroism, that the conformation of many rigid asymmetric polyamides
can be well-defined in solution. One of our most exciting discoveries has been the conformational transition of the polymers
derived from trans optically active dicarboxylic cyclopropanes and
diamines such as piperazine. In this example we show the first
convincing evidence of two rather well-controlled conformations
with a reasonably sharp transition between them, much like the
polyproline case. We have begun a systematic study of the effect
of substituents on the conformation of polyproline type polymers.
We can influence markedly the activation energy for the rotation
of the amide bond in polyprolines by appropriate substitution on
the pyrrolidine ring. By introducing a methyl substituent in the
2-position of proline, we can prepare a polymer which does not
mutarotate under ordinary conditions. We expect to continue
research in this area.

3.   The Use of Polymers as Reactants in Organic Reactions

We discovered some years ago that polymers derived from poly-
vinylimidazoles are catalysts for esterolytic reactions. A
systematic study of this type of system has revealed that this
catalytic activity can be attributed to three factors: 1) Cooperative effects, 2) electrostatic interactions, and 3) hydrophobic interactions. The concept of hydrophobic interactions
contributing to large acceleration of catalytic activity in ester
hydrolysis has provided us with a unique discovery which may be
applicable to many other kinds of organic reactions. This area
of chemistry has become much more popular in more recent years and
now there is an active group of investigators working in the
general field which has been expanded to all types of modifications
of these systems including micellar reactions. We believe we have
made a substantive original contribution to this area and intend
to pursue it with increased activity.

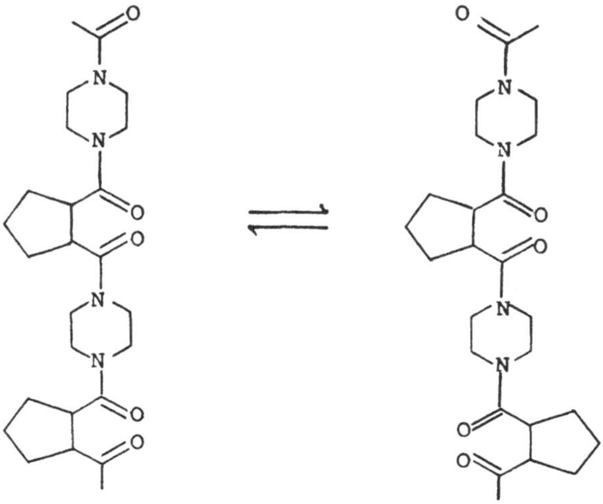

cisoid   transoid

Eq. 10

$$\text{PPI} \xrightleftharpoons[\text{Dilution of Acid Solutions with Alcohols}]{\text{Organic Acids}} \text{PPII}$$

Eq. 11

Table IX

Catalytic Rate Constants for the Solvolysis of PNPA
Catalyzed by Poly-4(5)-Vinylimidazole and Imidazole*
(Neutral-Neutral Imidazole Interactions)

pH	Poly-4(5)-Vinylimidazole		Imidazole	
	α_1	k_{cat}	α_1	k_{cat}
6.03	0.35	3.2	0.09	5.5
6.70	0.52	14.1	0.32	13.5
7.12	0.67	28.9	0.57	22.7
7.48	0.78	41.8	0.75	25.5

* In 10% methanol-water, ionic strength 0.02, k_{cat}
  in ℓ/mol/min.

$$k_{cat} = k_1\alpha_1 + k_2\alpha_1^2 + k_3\alpha_2 + k_4\alpha_1\alpha_2$$

4. Grafting of Electron Rich Substituents on the
    Side Chain of Linear Polymers

Our efforts in this area are concentrated currently on the
selective grafting of purine bases and amino acids onto linear
polyethyleneimine and related polymeric backbones. We are particularly interested in electron transfer through the side chains.

Adenin Unit   Thymin Unit

Table X

Neutral-Anionic Imidazole Interactions

X-1. Catalytic Rate Constants of Poly-N-Vinylimidazole
and Poly-2-Vinyl-N-methylimidazole with PNPA

	k_{cat}, ℓ/mole min	
pH	Poly-N-vinylimidazole	Poly-2-methyl-N-vinylimidazole
7.2	3.0	0.5
8.0	3.5	0.5
9.2	3.5	0.5

X-2. Catalytic Rate Constants of Imidazole and
Poly-4(5)-Vinylimidazole with NABA

	k_{cat}, ℓ/mole/min	
pH	Poly-4(5)-Vinylimidazole	Imidazole
6.1	35.8	11.3
7.2	120.2	24.4
8.2	104.4	31.9
9.0	55.5	34.5

Table XI

Influence of 1-Propanol-Water Mixtures on the Rate Constants (k_{cat}) for the PVIm and Im Catalyzed Hydrolyses of S_n^-

| Substrate | Catalyst | k_{cat} ($M^{-1}min^{-1}$) Vol. % 1-propanol-water[a] | | | | |
		15	20	25	30	40
S_2^- [b]	PVIm	297	145.5	89.5	59.5	34.2
	Im	65.5	44.5	30.0	173	10.2
	r [c]	4.6	3.3	3.0	3.4	3.4
S_7^-	PVIm	3,100	192.4	43.0	19.7	8.3
	Im	63.3	24.4	9.4	4.8	2.8
	r	49.0	8.0	4.6	4.1	3.0
S_{12}^-	PVIm	10,200	3,830	297	17.7	5.3
	Im	24.9	12.2	8.2	4.7	2.7
	r	410	314	36.3	3.8	2.0
S_{18}^-	PVIm		6,085	1,075	276	6.3
	Im		5.2	6.2	4.7	2.0
	r		1,170	172	58.9	2.3

[a] $[PVIm] = [Im] = 5.0 \times 10^{-4}\underline{M}$, $[S_n^-] = 5.0 \times 10^{-5}\underline{M}$, $\mu = 0.02$ $[Tris] = 0.02\underline{M}$, pH = 8.0, 26°.

[b]

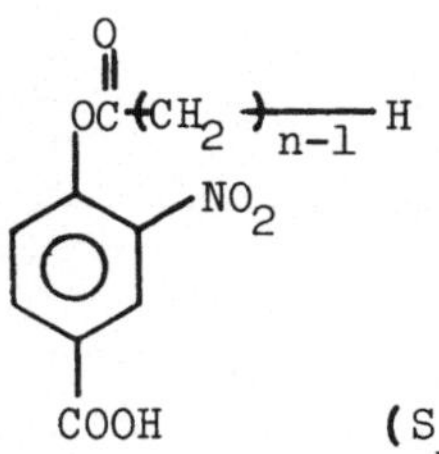

(S_n^-)

[c] $r = k_{cat}(PVIm)/k_{cat}(Im)$.

PUBLICATIONS OF DR. C. G. OVERBERGER

1. C. S. Marvel and C. G. Overberger, *J. Am. Chem. Soc.*, 66, 475 (1944).
2. C. S. Marvel and C. G. Overberger, ibid., 67, 2250 (1945).
3. C. S. Marvel and C. G. Overberger, ibid., 68, 185 (1946).
4. C. G. Overberger and R. E. Allen, ibid., 68, 722 (1946).
5. C. S. Marvel, C. G. Overberger, R. E. Allen, and J. H. Saunders, ibid., 68, 736 (1946).
6. C. S. Marvel, C. G. Overberger, R. E. Allen, H. W. Johnston, J. H. Saunders, and J. D. Young, ibid., 68, 861 (1946).
7. C. S. Marvel, J. H. Saunders, and C. G. Overberger, ibid., 68, 1085 (1946).
8. C. S. Marvel, R. E. Allen, and C. G. Overberger, ibid., 68, 1088 (1946).
9. C. S. Marvel and C. G. Overberger, ibid., 68, 2106 (1946).
10. A. C. Cope and C. G. Overberger, ibid., 69, 976 (1947).
11. A. C. Cope and C. G. Overberger, ibid., 70, 1433 (1948).
12. C. S. Marvel, R. Deanen, C. G. Overberger, and B. M. Kuhn, *J. Polymer Science*, 3, 128 (1948).
13. T. Alfrey, Jr., L. Arond, and C. G. Overberger, ibid., 4, 539 (1949).
14. C. G. Overberger and J. H. Saunders, <u>Organic Syntheses</u>, Vol. 28, John Wiley and Sons, Inc., 1948, p. 31.
15. C. G. Overberger, J. H. Saunders, R. E. Allen, and R. Gander, ibid., p. 28.
16. C. G. Overberger, E. Luhrs, and P. K. Chien, *J. Am. Chem. Soc.*, 72, 1200 (1950).
17. C. G. Overberger, S. P. Ligthelm, and E. A. Swire, ibid., 72, 2856 (1950).
18. C. G. Overberger, H. J. Mallon, and R. Fine, ibid., 72, 4958 (1950).
19. C. G. Overberger and H. Mark, <u>Chemical Industries</u>, Aug. 1950, p. 192.
20. C. G. Overberger and C. W. Roberts, *J. Am. Chem. Soc.*, 71, 3618 (1949).
21. C. G. Overberger, M. T. O'Shaughnessy, and H. Shalit, ibid., 71, 2661 (1949).
22. C. G. Overberger and John M. Hoyt, ibid., 73, 3305 (1951).
23. C. G. Overberger and J. M. Hoyt, ibid., 73, 3957 (1951).
24. C. G. Overberger and J. Lal, ibid., 73, 2956 (1951).
25. C. G. Overberger, Paul Fram, and T. Alfrey, Jr., *J. Polymer Sci.*, 6, 539 (1951).
26. C. G. Overberger and M. B. Berenbaum, *J. Am. Chem. Soc.*, 73, 2618 (1951).
27. C. G. Overberger, A. Fischman, C. W. Roberts, L. H. Arnold, and J. Lal, ibid., 73, 2540 (1951).
28. A. C. Cope, H. L. Dryden, Jr., C. G. Overberger, and A. A. D'Addieco, ibid., 73, 3416 (1951).

29. C. G. Overberger, D. E. Baldwin, and H. P. Gregor, ibid.,
     72, 4864 (1950).

30. H. P. Gregor, J. I. Bregman, V. Gutoff, R. D. Broadly,
     D. E. Baldwin, and C. G. Overberger, *J. Colloid Sci.*, 6,
     20 (1951).

31. C. G. Overberger, L. H. Arond, R. H. Wiley, and R. R. Garrett,
     J. Polymer Sci., 7, 4 (1951).

32. C. G. Overberger and H. Biletch, *J. Am. Chem. Soc.*, 73, 4880
     (1951).

33. C. G. Overberger and M. B. Berenbaum, ibid., 73, 4883 (1951).

34. C. G. Overberger, T. B. Gibb, Jr., S. Chibnik, Pao-tung Huang,
     and J. J. Monagle, ibid., 74, 3290 (1952).

35. C. G. Overberger and M. B. Berenbaum, ibid., 74, 3293 (1952).

36. C. G. Overberger, L. H. Arond, and J. J. Taylor, ibid., 73,
     5541 (1951).

37. C. G. Overberger, J. J. Taylor, and T. Alfrey, Jr., ibid.,
     74, 4848 (1952).

38. C. G. Overberger, P. T. Huang, and M. B. Berenbaum, Organic
     Syntheses, Vol. 32, John Wiley & Sons, Inc., 1952, p. 16.

39. C. G. Overberger and M. B. Berenbaum, ibid., p. 48.

40. C. G. Overberger, P. T. Huang, and M. B. Berenbaum, ibid.,
     p. 50.

41. C. G. Overberger, Morton A. Klotz, and H. Mark, *J. Am. Chem.
     Soc.*, 75, 3186 (1953).

42. C. J. Johnson, C. G. Overberger, and W. J. Seagers, ibid.,
     75, 1495 (1953).

43. C. G. Overberger, Ramon A. Gadea, Jane Ann Smith, and
     Irving C. Kogon, ibid., 75, 2075 (1953).

44. C. G. Overberger, Harry Biletch, A. B. Finestone, Jay Lilker,
     and James Herbert, ibid., 75, 2078 (1953).

45. C. G. Overberger, Pao-tung Huang, and Thomas B. Gibb, Jr.,
     ibid., 75, 2082 (1953).

46. C. G. Overberger, Charles Frazier, Jerome Mandelman, and
     Harry F. Smith, ibid., 75, 3326 (1953).

47. C. G. Overberger and R. W. Cummins, ibid., 75, 4250 (1953).

48. C. G. Overberger and R. W. Cummins, ibid., 75, 4783 (1953).

49. C. G. Overberger and Peter Kabasakalian, ibid., 75, 6058
     (1953).

50. C. G. Overberger and Gerard F. Endres, ibid., 75, 6349 (1953).

51. C. G. Overberger and Seymour L. Shapiro, ibid., 76, 97 (1954).

52. C. G. Overberger and Seymour L. Shapiro, ibid., 76, 93 (1954).

53. C. G. Overberger, R. J. Ehrig, and David Tanner, ibid., 76,
     772 (1954).

54. C. G. Overberger and Seymour L. Shapiro, ibid., 76, 1061
     (1954).

55. C. G. Overberger and Irving C. Kogon, ibid., 76, 1065.

56. C. G. Overberger and Seymour L. Shapiro, ibid., 76, 1855
     (1954).

57. C. G. Overberger and Irving C. Kogon, ibid., 76, 1879 (1954).

58. C. G. Overberger, Irving C. Kogon, and W. J. Einstein, ibid., 76, 1953 (1954).

59. C. G. Overberger and Alexander Lebovits, ibid., 76, 2722 (1954).

60. C. G. Overberger, Warren F. Hale, M. B. Berenbaum, and A. B. Finestone, ibid., 76, 6185 (1954).

61. C. G. Overberger, P. M. Kamath, and I. Tashlick, J. Org. Chem., 19, 1486 (1954).

62. C. G. Overberger and David Tanner, J. Am. Chem. Soc., 77, 369 (1955).

63. C. G. Overberger and Gerard F. Endres, ibid., 77, 2201 (1955).

64. C. G. Overberger and Alexander Lebovits, ibid., 77, 3675 (1955).

65. C. G. Overberger and Burton S. Marks, ibid., 77, 4097 (1955).

66. C. G. Overberger, Louis C. Palmer, Burton S. Marks, and Norman R. Byrd, ibid., 77, 4100 (1955).

67. C. G. Overberger, B. S. Marks, ibid., 77, 4101 (1955).

68. C. G. Overberger and Milton Lapkin, ibid., 77, 4651 (1955).

69. C. G. Overberger and Gerard F. Endres, J. Polymer Sci., 16, 283 (1953).

70. Irving C. Kogon, Ronald Minin, and C. G. Overberger, Organic Syntheses, Vol. 35, John Wiley & Sons, Inc., 1955. p. 34.

71. C. G. Overberger and Louis C. Palmer, J. Am. Chem. Soc., 78, 666 (1956).

72. C. G. Overberger and A. B. Finestone, ibid., 78, 1638 (1956).

73. C. G. Overberger, Harry Biletch, Pao-tung Huang, and H. M. Blatter, J. Org. Chem., 20, 1717 (1955).

74. C. G. Overberger, Norman R. Byrd, and R. B. Mesrobian, J. Am. Chem. Soc., 78, 1961 (1956).

75. C. G. Overberger and A. Katchman, ibid., 78, 1965 (1956).

76. C. G. Overberger, G. F. Endres, and Avito Monaci, ibid., 78, 1969 (1956).

77. C. G. Overberger and John J. Monagle, ibid., 78, 4470 (1956).

78. C. G. Overberger and Alexander Lebovits, ibid., 78, 4792 (1956).

79. C. G. Overberger and Peter Kabasakalian, J. Org. Chem., 21, 1124 (1956).

80. C. G. Overberger, J. E. Mulvaney, and F. Marshall Beringer, ibid., 21, 1311 (1956).

81. C. G. Overberger and George P. Schimitt, J. Am. Chem. Soc., 79, 508 (1957).

82. C. G. Overberger, F. W. Michelotti, and P. M. Carabateas, ibid., 79, 941 (1957).

83. C. G. Overberger, Joseph G. Lombardino, and Richard G. Hiskey, ibid., 79, 1510 (1957).

84. C. G. Overberger, Irving Tashlick and Richard G. Hiskey, ibid., 79, 2662 (1957).

85. C. G. Overberger and Peter Kabasakalian, ibid., 79, 3182 (1957).

86.   C. G. Overberger, Joseph G. Lombardino, and Richard G. Hiskey, *J. Org. Chem.*, <u>22</u>, 858 (1957).

87.   C. G. Overberger, U.S. Patent 2,802,812 - 8/13/57.

88.   C. G. Overberger, Joseph G. Lombardino, and Richard G. Hiskey, *J. Am. Chem. Soc.*, <u>79</u>, 6430 (1957).

89.   C. G. Overberger, H. Biletch, and R. G. Nickerson, *J. Polymer Sci.*, <u>27</u>, (issue 115), p. 381 (1958).

90.   C. G. Overberger and Francis W. Michelotti, *J. Am. Chem. Soc.*, <u>80</u>, 988 (1958).

91.   C. G. Overberger, Eli M. Pearce, and D. Tanner, ibid., <u>80</u>, 1761 (1958).

92.   C. G. Overberger and Joseph G. Lombardino, ibid., <u>80</u>, 2317 (1958).

93.   C. G. Overberger, R. J. Ehrig, and R. A. Marcus, ibid., <u>80</u>, 2456 (1958).

94.   C. G. Overberger, Joseph G. Lombardino, and Richard G. Hiskey, ibid., <u>80</u>, 3009 (1958).

95.   C. G. Overberger, Eli M. Pearce, and L. C. Kung Wei, *J. Polymer Sci.*, <u>31</u> (issue 122), p. 235 (1958).

96.   C. G. Overberger, Eli M. Pearce, and N. Mayes, ibid., <u>31</u> (issue 122), p. 217 (1958).

97.   C. G. Overberger and Harold Gainer, *J. Am. Chem. Soc.*, <u>80</u>, 4556 (1958).

98.   C. G. Overberger and Harold Gainer, ibid., <u>80</u>, 4561 (1958).

99.   C. G. Overberger, D. Tanner, and Eli M. Pearce, ibid., <u>80</u>, 4566 (1958).

100.   C. G. Overberger and P. V. Bonsignore, ibid., <u>80</u>, 5431 (1958).

101.   C. G. Overberger and P. V. Bonsignore, ibid., <u>80</u>, 5427 (1958).

102.   C. G. Overberger and Arthur Katchman, *Chemical & Engineering News*, Vol. 36, p. 80 (November 10, 1958).

103.   C. G. Overberger, <u>The Chemist</u> (American Institute of Chemists) Vol. 35, No. 12, p. 519.

104.   C. G. Overberger, I. Tashlick, M. Bernstein, and R. G. Hiskey, *J. Am. Chem. Soc.*, <u>80</u>, 6556 (1958).

105.   C. G. Overberger and Adolph V. DiGiulio, ibid., <u>80</u>, 6562 (1958).

106.   C. G. Overberger and S. H. Pinner, ibid., <u>75</u>, 4221 (1953).

107.   C. G. Overberger and I. Tashlick, *J. Am. Chem. Soc.*, <u>81</u>, 217, (1959).

108.   C. G. Overberger, Eli M. Pearce, and N. Mayes, *J. Polymer Science*, Vol. 34, p. 109-120 (1959).

109.   C. G. Overberger and Aino Lusi, *J. Am. Chem. Soc.*, <u>81</u>, 506 (1959).

110.   C. G. Overberger, F. Ang, and H. Mark, *J. Polymer Science*, Vol. 35, 381-389 (1959).

111.   C. G. Overberger, H. Biletch, and F. W. Orttung, *J. Org. Chem.*, <u>24</u>, 289 (1959).

112.   C. G. Overberger and A. V. DiGiulio, *J. Am. Chem. Soc.*, <u>81</u>, 1194 (1959).

113. C. G. Overberger, N. P. Marullo, and R. G. Hiskey, ibid., 81, 1517 (1959).
114. C. G. Overberger and A. V. DiGiulio, ibid., 81 2154 (1959).
115. C. G. Overberger and V. G. Kamath, ibid., 81, 2910 (1959).
116. C. G. Overberger, George Kesslin, and Pao-tung Huang, ibid., 81, 3779 (1959).
117. C. G. Overberger and John M. Whelan, J. Org. Chem., 24, 1155 (1959).
118. C. G. Overberger and J. E. Mulvaney, J. Am. Chem. Soc., 81, 4697 (1959).
119. C. G. Overberger, Michael P. Mazzeo, and J. J. Godfrey, J. Org. Chem., 24, 1407 (1959).
120. C. G. Overberger and J. J. Godfrey, J. Polymer Science, Vol. 40, p. 179-201 (1959).
121. Arnold J. Rosenthal and C. G. Overberger, J. Am. Chem. Soc., 82, 108 (1960).
122. C. G. Overberger and A. J. Rosenthal, ibid., 82, 117 (1960).
123. C. G. Overberger and F. Ang, ibid., 82, 929 (1960).
124. C. G. Overberger and A. E. Borchert, ibid., 82, 1007 (1960).
125. C. G. Overberger, A. Katchman, and J. E. Mulvaney, J. Org. Chemistry, 25, 270 (1960).
126. C. G. Overberger, Record of Chemical Progress, Vol. 21, No. 1 (1960).
127. A. E. Borchert and C. G. Overberger, J. Polymer Sci., Vol. 44, p. 483 (1960).
128. A. E. Borchert, C. G. Overberger, and A. Katchman, ibid., 44, p. 491 (1960).
129. C. G. Overberger and M. G. Newton, J. Am. Chem. Soc., 82, 3622 (1960).
130. C. G. Overberger and Herbert Aschkenasy, ibid., 82, 4357 (1960).
131. C. G. Overberger and Herbert Aschkenasy, J. Org. Chemistry, 25, 1648 (1960).
132. C. G. Overberger and A. E. Borchert, J. Am. Chem. Soc., 82, 4896 (1960).
133. C. G. Overberger, H. Yuki, and N. Urakawa, J. Pol. Science, Vol. 45, p. 127 (1960).
134. C. G. Overberger and Pao-kung Chien, J. Am. Chem. Soc., 82, 5874 (1960).
135. C. G. Overberger, Transactions of The New York Academy of Sciences, Vol. 23, No. 1 (1960), p. 25.
136. C. G. Overberger, J. Pol. Sci., Vol. XVLII, p. 533, (1960).
137. C. G. Overberger, N. P. Marullo, and R. G. Hiskey, J. Am. Chem. Soc., 83, 1374 (1961).
138. C. G. Overberger and N. P. Marullo, ibid., 83, 1378 (1961).
139. Helmut Ringsdorf and C. G. Overberger, Die Makromolekulare Chemie, XLIV-XLVI, 418 (1961).
140. Reed F. Riley and C. G. Overberger, Journal of Chemical Education, Vol. 38, 424 (1961).

141. C. G. Overberger, J. *Polymer Sci.*, Vol. 51, p. S-77
      (1961).
142. C. G. Overberger, J. J. Ferraro, and F. W. Orttung, J. *Org.
      Chem.*, 26, 3458 (1961).
143. C. G. Overberger and Arthur M. Schiller, J. *Pol. Sci.*,
      Vol. 54, p. S-30 (1961).
144. C. G. Overberger and Arthur M. Schiller, J. *Org. Chem.*,
      26, 4230 (1961).
145. C. G. Overberger and John M. Whelan, ibid., 26, 4328 (1961).
146. C. G. Overberger and J. Richard Hall, ibid., 26, 4359 (1961).
147. C. G. Overberger and E. Sarlo, ibid., 26, 4711 (1961).
148. J. E. Mulvaney, C. G. Overberger, and Arthur M. Schiller,
      Fortschritte Der Hochpolymeren-Forschung (*Advances in
      Polymer Science*), 3, Band, 1 Heft, s. 106 (1961).
149. C. G. Overberger and H. Jabloner, J. *Pol. Sci.*, 55, S-32
      (1961).
150. C. G. Overberger and Louis P. Herin, J. *Org. Chem.*, 27,
      417 (1962).
151. C. G. Overberger and Jean-Pierre Anselme, J. *Am. Chem. Soc.*,
      84, 869 (1962).
152. C. G. Overberger, G. Kesslin, and N. R. Byrd, J. *Org. Chem.*,
      27, 1568 (1962).
153. C. G. Overberger, J. J. Ferraro, P. V. Bonsignore, F. W.
      Orttung, and N. Vorchheimer, *Pure and Applied Chemistry*,
      Vol. 4, 521 (1962).
154. C. G. Overberger, J. *Org. Chem.*, 27, 2267 (1962).
155. C. G. Overberger, Paul Barkan, Aino Lusi, and Helmut
      Ringsdorf, J. *Amer. Chem. Soc.*, 84, 2814 (1962).
156. Helmut Ringsdorf and C. G. Overberger, J. *Pol. Sci.*, Vol. 61,
      Issue 171, S-11 (1962).
157. C. G. Overberger and E. B. Davidson, ibid., Vol. 62, p. 23-
      31 (1962).
158. C. G. Overberger, S. Ishida, and H. Ringsdorf, ibid.,
      Vol. 62, Issue 173 (1962).
159. C. G. Overberger and L. P. Herin, J. *Org. Chem.*, 27, 2423
      (1962).
160. C. G. Overberger and J. J. Ferraro, ibid., 27, 3539 (1962).
161. C. G. Overberger and G. Kesslin, ibid., 27, 3898 (1962).
162. C. G. Overberger, H. Ringsdorf, and N. Weinshenker, ibid.,
      27, 4331 (1962).
163. G. F. Endres, V. G. Kamath, and C. G. Overberger, J. *Am.
      Chem. Soc.*, 84, 4813 (1962).
164. C. G. Overberger and Jean-Pierre Anselme, J. *Org. Chem.*, 28,
      592 (1963).
165. C. G. Overberger and V. G. Kamath, J. *Amer. Chem. Soc.*, 85,
      446 (1963).
166. C. G. Overberger, Martin Tobkes, and Arnold Zweig, J. *Org.
      Chem.*, 28, 620 (1963).
167. C. G. Overberger and George Walter Halek, ibid., 28, 867
      (1963).

168. Carl Djerassi, B. F. Burrows, C. G. Overberger, T. Takekoshi, C. D. Gutsche, and C. T. Chang, *J. Amer. Chem. Soc.*, 85, 949 (1963).

169. C. G. Overberger and N. Vorchheimer, ibid., 85, 951 (1963).

170. C. G. Overberger and Jean-Pierre Anselme, *Tetrahedron Letters No. 21*, pp. 1405-08 (1963).

171. C. G. Overberger and Peter A. Jarovitzky, *J. Polymer Sci.*, Part C, No. 4, p. 37-48 (1964).

172. C. G. Overberger and A. Drucker, *J. Org. Chem.*, 29, 360 (1964).

173. C. G. Overberger and Jean-Pierre Anselme, *J. Am. Chem. Soc.*, 86, 658 (1964).

174. C. G. Overberger, H. Kaye, and G. Walsh, *J. Polymer Sci.*, Part A, 2, 755-767 (1964).

175. C. G. Overberger and E. Sarlo, ibid., Part A, 2, pp. 1017-1021 (1964).

176. C. G. Overberger and Jean-Pierre Anselme, Chemistry and Industry, 1964, pp. 280-281.

177. C. G. Overberger and W. H. Daly, *J. Org. Chem.*, 29, 757 (1964).

178. C. G. Overberger and Jurgen Weise, *J. Polymer Sci.*, (Polymer Letters), 2, pp. 329-331 (1964).

179. C. G. Overberger and Jean-Pierre Anselme, *J. Org. Chem.*, 29, 1188 (1964).

180. C. G. Overberger and Martin Tobkes, *J. Polymer Sci.*, Part A, 2, pp. 2481-2487 (1964).

181. C. G. Overberger, S. Ozaki, and H. Mukamal, *J. Polymer Sci.*, (Polymer Letters), 2, p. 627 (1964).

182. C. G. Overberger and Herbert A. Friedman, *J. Org. Chem.*, 29, 1720 (1964).

183. C. G. Overberger and W. H. Daly, *J. Am. Chem. Soc.* 86, 3402 (1964).

184. The Chemistry of Cationic Polymerization (book by P. H. Plesch). Book review by C. G. Overberger, ibid., 86, 4739 (1964).

185. C. G. Overberger, N. Weinshenker, J. P. Anselme, ibid., 86, 5364 (1964).

186. C. G. Overberger, F. S. Diachkovsky, and P. A. Jarovitzky, *J. Polymer Sci.*, Part A, 2, p. 4113 (1964).

187. C. G. Overberger, H. Ringsdorf, and B. Avchen, *J. Org. Chem.*, 30, 232 (1965).

188. C. G. Overberger, T. St. Pierre, N. Vorchheimer, J. Lee, and S. Yaroslavsky, *J. Am. Chem. Soc.*, 87, 296 (1965).

189. C. G. Overberger and P. A. Jarovitzky, *J. Polymer Sci.*, Part A, 3, p. 1483 (1965).

190. C. G. Overberger, J. E. Mulvaney, and Arthur M. Schiller, Chapter in Volume 2 of Encyclopedia of Polymer Science and Technology, pp. 95-137, Interscience - Wiley, New York, NY (1965).

191. C. G. Overberger and Herbert A. Friedman, *J. Org. Chem.*, 30, 1926 (1965).
192. C. G. Overberger and Naoki Yamamoto, *J. Polymer Sci.*, (*Polymer Letters*), 3, p. 569 (1965).
193. C. G. Overberger, R. Sitaramaiah, T. St. Pierre, and S. Yaroslavsky, *J. Am. Chem. Soc.*, 87, 3270 (1965).
194. C. G. Overberger, T. St. Pierre, N. Vorchheimer, C. Yaroslavsky, J. Lee, J. C. Salamone, R. Sitaramaiah, S. Yaroslavsky, *Transactions of the New York Academy of Sciences*, Vol. 27, No. 7, p. 790 (1965).
195. C. G. Overberger, T. M. Chapman, and T. Wojnarowski, *J. Polymer Sci.*, Part A, 3, p. 2865 (1965).
196. C. G. Overberger and Sumio Fujimoto, ibid., Part B, 3, p. 735 (1965).
197. C. G. Overberger and Susumu Ishida, ibid., Part B, 3, p. 789 (1965).
198. C. G. Overberger, H. Ringsdorf, and B. Avchen, *J. Org. Chem.*, 30, 3088 (1965).
199. C. G. Overberger, N. Weinschenker, and Jean-Pierre Anselme, *J. Amer. Chem. Soc.*, 87, 4119 (1965).
200. C. G. Overberger, K. H. Burg, and W. H. Daly, ibid., 87, 4125 (1965).
201. C. G. Overberger, T. St. Pierre, and S. Yaroslavsky, ibid., 87, 4310 (1965).
202. C. G. Overberger, J. C. Salamone, and S. Yaroslavsky, *J. Org. Chem.*, 30, 3580 (1965).
203. C. G. Overberger and Herbert A. Friedman, *J. Polymer Sci.*, Part A, 3, p. 3625 (1965).
204. C. G. Overberger, H. Ringsdorf, and B. Avchen, *J. Med. Chem.*, 8, 862 (1965).
205. C. G. Overberger, T. Kurtz, and S. Yaroslavsky, *J. Org. Chem.*, 30, 4363 (1965).
206. C. G. Overberger and Carmela Yaroslavsky, *Tetrahedron Letters*, No. 49, p. 4395 (1965).
207. C. G. Overberger and T. Kurtz, *J. Org. Chem.*, 31, 288 (1966).
208. C. G. Overberger, T. St. Pierre, C. Yaroslavsky, and S. Yaroslavsky, *J. Am. Chem. Soc.*, 88, 1184 (1966).
209. C. G. Overberger and Peter A. Jarovitzky, *J. Polymer Sci.*, Part C, No. 12, p. 3-21 (1966).
210. C. G. Overberger, S. Ozaki, and D. M. Braunstein, *Die Makromolekulare Chemie*, Band 93, Seite 13 (1966).
211. C. G. Overberger and Siegfried Altscher, *J. Org. Chem.*, 31, 1728 (1966).
212. C. G. Overberger, Richard E. Zangaro, and J.-P. Anselme, ibid., 31, 2046 (1966).
213. C. G. Overberger, J.-P. Anselme, and J. G. Lombardino, Organic Compounds with Nitrogen-Nitrogen Bonds, The Ronald Press Company, New York, NY, August, 1966.
214. H. Morawetz, C. G. Overberger, J. C. Salamone, and S. Yaroslavsky, *J. Polymer Sci.*, Part B, 4, p. 609 (1966).

215. A. Simon, P. A. Jarovitzky, and C. G. Overberger, ibid., Part A-1, 4, p. 2513 (1966).

216. C. G. Overberger and Naoki Yamamoto, ibid., Part A-1, 4, p. 3101 (1966).

217. C. G. Overberger, B. Kösters, and T. St. Pierre, J. Polymer Sci., Part A-1, Vol. 5, 1987-92 (1967).

218. C. G. Overberger, Science, Vol. 158, No. 3797, 109 (1967).

219. C. G. Overberger, Presidential Address presented Sept. 11, 1967, at 154th ACS National Meeting in Chicago, C. and E. News, Vol. 45, 88-92 (1967).

220. (A.) C. G. Overberger, P. A. Jarovitzky, and H. Mukamal, J. Polymer Sci., Part A-1, Vol. 5, 2487-2502 (1967). (B.) C. G. Overberger and Howard Kaye, J. Am. Chem. Soc., 89, 5640-5645 (1967).

221. C. G. Overberger and Howard Kaye, J. Am. Chem. Soc., 89, 5646-5649 (1967).

222. C. G. Overberger and Howard Kaye, J. Am. Chem. Soc., 89, 5649-5655 (1967).

223. C. G. Overberger, J. C. Salamone, and S. Yaroslavsky, J. Am. Chem. Soc., 89, 6231-6236 (1967).

224. C. G. Overberger and Iwhan Cho, J. of Franklin Inst., Vol. 284, 432, Dec., 1967.

225. C. G. Overberger, J. C. Salamone, and S. Yaroslavsky, Israel J. Chem. Proceedings, Vol. 5, p. 66 (1967).

226. C. G. Overberger, H. Morawetz, J. C. Salamone, and S. Yaroslavsky, J. Am. Chem. Soc., 90, 651 (1968).

227. C. G. Overberger and T. Takekoshi, Macromolecules, Vol. 1, 7 (1968).

228. C. G. Overberger, J. C. Salamone, and J. Sebenda, Science, Vol. 159, 1224 (1968).

229. C. G. Overberger and T. Takekoshi, Macromolecules, Vol. 1, 1 (1968).

230. C. G. Overberger and Gordon M. Parker, J. of Polymer Sci., Part A-1, Vol. 6, 513-526 (1968).

231. C. G. Overberger, J. C. Salamone, and S. Yaroslavsky, Pure and Applied Chem., Vol. 15, 453-464 (1967).

232. C. G. Overberger and J. A. Moore, Encyclopedia of Polymer Science and Tech., Vol. 7, 743-753 (1967).

233. C. G. Overberger and Gordon M. Parker, J. Polymer Sci., Part C, No. 22, pp. 387-406 (1968).

234. C. G. Overberger and Jurgen K. Weise, J. Am. Chem. Soc., 90, 3538 (1968).

235. C. G. Overberger and Jurgen K. Weise, J. Am. Chem. Soc., 90, 3525 (1968).

236. C. G. Overberger and Jurgen K. Weise, J. Am. Chem. Soc., 90, 3533 (1968).

237. C. G. Overberger and J. A. Moore, The Chemistry of Sulfides, edited by Arthur V. Tobolsky, Interscience Publishing, Div. of John Wiley & Sons, New York, NY (1968).

238. C. G. Overberger, R. Corett, J. C. Salamone, and
S. Yaroslavsky, *Macromolecules*, Vol. 1, p. 331, (1968).

239. C. G. Overberger and Iwhan Cho, *J. Org. Chem.*, 33, 3321
(1968).

240. C. G. Overberger and J. A. Moore, *Chem. and Ind.*, p. 24
(1968).

241. C. G. Overberger, *J. of Chem. Educ.*, 45, p. 502 (August,
1968).

242. C. G. Overberger and Iwhan Cho, *J. of Polymer Sci.*, Part A-1,
6, 2741-2754 (1968).

243. C. G. Overberger, H. Maki, and J. C. Salamone, *Svensk
Kemisk Tidskrift*, 80:5 (1968).

244. C. G. Overberger and C. M. Shen, *Organic Preparations and
Procedures*, 1 (1), 1-3 (1969).

245. C. G. Overberger, J. C. Salamone, I. Cho, and H. Maki,
Annals of the New York Academy of Sciences, Vol. 155,
Article 2, pp. 431-446 (1969).

246. C. G. Overberger and Sumio Fujimoto, *J. of Polymer Sci.*,
Part C, No. 16, pp. 4141-4168 (1968).

247. C. G. Overberger and G. Montaudo, *J. Am. Chem. Soc.*, 91,
753 (1969).

248. C. G. Overberger, M. Valentine, and J.-P. Anselme, *J. Am.
Chem. Soc.*, 91, 687 (1969).

249. C. G. Overberger, G. Montaudo, J. Sebenda, and R. A. Veneski,
J. Am. Chem. Soc., 91, 1256 (1969).

250. C. G. Overberger, A. Wartman, and J. C. Salamone, *Organic
Preparations and Procedures*, 1 (2), 117 (1969).

251. C. G. Overberger, *R.G.C.P.*, 46, n° 4 (1969).

252. C. G. Overberger and G. Khattab, *J. of Polymer Sci.*, Part
A-1, 7, 217 (1969).

253. C. G. Overberger, G. Montaudo, and S. Ishida, *J. of Polymer
Sci.*, Part A-1, 7, 35 (1969).

254. C. G. Overberger, H. D. Noether, and G. Halek, *J. of Polymer
Sci.*, Part A-1, 7, 201 (1969).

255. C. G. Overberger and C. M. Burns, *J. Polymer Sci.*, Part A-1,
7, 333 (1969).

256. C. G. Overberger, C. Yaroslavsky, Hyman Katz, J.-P. Anselme,
and J. W. Stoddard, *J. Amer. Chem. Soc.*, 91, 311 (1969).

257. C. G. Overberger, G. Montaudo, M. Mizutori, J. Sebenda, and
R. A. Veneski, *J. Polymer Sci.*, Part B, 7, 225 (1969).

258. C. G. Overberger, G. Montaudo, Y. Nishimura, J. Sebenda, and
R. A. Veneski, *J. Polymer Sci.*, Part B, 7, 219 (1969).

259. C. G. Overberger and Siegfried Altscher, *J. Chemical and
Engineering Data*, 14, 266 (1969).

260. C. G. Overberger and J. C. Salamone, *Accounts of Chem. Res.*,
2, 217 (1969).

261. C. G. Overberger and Tohru Takekoshi, *J. Polymer Sci.*, Part
A-1, 7, 1011 (1969).

262. C. G. Overberger, M. Morimoto, I. Cho, and J. C. Salamone,
Macromolecules, 2, 553 (1969).

263. C. G. Overberger and Jan Sebenda, *J. Polymer Sci.*, A-1, 7, 2875 (1969).

264. C. G. Overberger, Jules Reichenthal, and J.-P. Anselme, *J. Org. Chem.*, 35, 138 (1970).

265. C. G. Overberger and J. A. Moore, *Adv. Polymer Sci.*, 7, 113 (1970).

266. C. G. Overberger, R. A. Veneski, and G. Montaudo, *J. Polymer Sci.*, Part B, 7, 877 (1969).

267. C. G. Overberger and P. S. Yuen, *J. Amer. Chem. Soc.*, 92, 1667 (1970).

268. C. G. Overberger and G. W. Halek, *J. Polymer Sci.*, A-1, 8, 359 (1970).

269. C. G. Overberger and H. Maki, *Macromolecules*, 3, 214 (1970).

270. C. G. Overberger and H. Maki, *Macromolecules*, 3, 220 (1970).

271. C. G. Overberger, *J. Rubb. Res. Inst. Malaya*, 22(2), 201-213 (1969).

272. C. G. Overberger and D. A. Labianca, *J. Org. Chem.*, 35, 1762 (1970).

273. C. G. Overberger, T. Yoshimura, A. Ohnishi, and A. S. Gomes, *J. Polymer Sci.*, A-1, 8, 2275-2291 (1970).

274. C. G. Overberger and John Stoddard, *J. Amer. Chem. Soc.*, 92, 4922 (1970).

275. C. G. Overberger and J. C. Salamone, *Buletinul Institutului Politehnic Din Iasi, Tomul XVI (XX)*, fasc. 1-2 (1970).

276. C. G. Overberger, G. Montaudo, Toshiuki Furuyama, and Murray Goodman, *J. Polymer Sci.*, Part C, 31, 33 (1970).

277. C. G. Overberger, Richard E. Zangaro, Rudolph E. K. Winter, and J.-P. Anselme, *J. Org. Chem.*, 36, 975 (1971).

278. C. G. Overberger and Y. Shimokawa, *Polymer Letters*, 9, 161 (1971).

279. C. G. Overberger, R. A. Veneski, and Jan Sebenda, *J. Polymer Sci.*, A-1, 9, 701 (1971).

280. C. G. Overberger, A. Ohnishi, and A. S. Gomes, *J. Polymer Sci.*, A-1, 9, 1139 (1971).

281. C. G. Overberger and M. Morimoto, *J. Amer. Chem. Soc.*, 93, 3222 (1971).

282. C. G. Overberger, M. Morimoto, I. Cho, and J. C. Salamone, *J. Amer. Chem. Soc.*, 93, 3228 (1971).

283. C. G. Overberger and Chah-Moh Shen, *Bioorganic Chemistry*, 1, 1 (1971).

284. C. G. Overberger and I. Scheinfeld, *J. Polymer Sci.*, A-1, 9, 3233 (1971).

285. C. G. Overberger and Y. Shimokawa, *Macromolecules*, 4, 718 (1971).

286. C. G. Overberger and Chah-Moh Shen, *J. Amer. Chem. Soc.*, 93, 6992 (1971).

287. C. G. Overberger, G. Montaudo, Y. Nishimura, J. Sebenda, and R. A. Veneski, *IUPAC International Symposium on Macromolecular Chemistry, Budapest*, 1969, pp. 127-153.

288. G. Montaudo, P. Maravigna, P. Finocchiaro, and C. G. Overberger, *Macromolecules*, 5, 203 (1972).
289. G. Montaudo, P. Finocchiaro, P. Maravigna, and C. G. Overberger, *Macromolecules*, 5, 197 (1972).
290. G. Montaudo, C. G. Overberger, *Polymer Letters*, 10, 433 (1972).
291. C. G. Overberger, K.-H. David, and J. A. Moore, *Macromolecules*, 5, 368 (1972).
292. C. G. Overberger and Y. Okamoto, *Macromolecules*, 5, 363 (1972).
293. C. G. Overberger and K.-H. David, *Macromolecules*, 5, 373 (1972).
294. C. G. Overberger, John H. Kozlowski, and Edward Radlmann, *J. Polymer Sci.*, A-1, 10, 2265 (1972).
295. C. G. Overberger and John H. Kozlowski, *J. Polymer Sci.*, A-1, 10, 2291 (1972).
296. C. G. Overberger, *J. Macromol. Sci.--Chem.*, A6(5), 855 (1972).
297. C. G. Overberger, M. S. Chi, Donald G. Pucci, and J. A. Barry, *Tetrahedron Lett.*, 45, 4565 (1972).
298. Y. Okamoto and C. G. Overberger, *J. Polymer Sci., Polymer Chemistry*, 10, 3387 (1972).
299. G. Montaudo and C. G. Overberger, *J. Org. Chem.*, 38, 804 (1973).
300. C. G. Overberger, R. C. Glowaky, and P.-H. Vandewyer, *J. Am. Chem. Soc.*, 95, 6008 (1973).
301. C. G. Overberger and R. C. Glowaky, *J. Amer. Chem. Soc.*, 95, 6014 (1973).
302. C. G. Overberger, G. Montaudo, R. C. Glowaky, and M. J. Han, *Macromolecular Chemistry--8*, IUPAC, International Symposium, Helsinki, 1972.
303. C. G. Overberger and K. Gerberding, *J. Polymer Sci., Polymer Letters*, 11, 465 (1973).
304. G. Montaudo, P. Finocchiaro, and C. G. Overberger, *J. Polymer Sci., Polymer Letters*, 11, 619 (1973).
305. G. Montaudo and C. G. Overberger, *J. Polymer Sci., Polymer Letters*, 11, 625 (1973).
306. G. Montaudo, P. Finocchiaro, and C. G. Overberger, *J. Polymer Sci., Polymer Chemistry*, 11, 2727 (1973).
307. G. Montaudo and C. G. Overberger, *J. Polymer Sci., Polymer Chemistry*, 11, 2739 (1973).
308. C. G. Overberger, *Appl. Polym. Symposium*, 22, 287 (1973).
309. C. G. Overberger and Thomas W. Smith, <u>Catalysis by Polymers</u>, edited by J. A. Moore, D. Reidel Publishing Company, Boston, 1973.
310. C. G. Overberger and C. J. Podsiadly, *Bioorganic Chemistry*, 3, 16 (1974).
311. C. G. Overberger and C. J. Podsiadly, *Bioorganic Chemistry*, 3, 35 (1974).

312. C. G. Overberger and K. N. Sannes, *Angew. Chem. Internat. Edit.*, 13, 99 (1974); *Angew. Chemie*, 86, 139 (1974); *Intern. J. Polymeric Mater.*, Vol. 2, 271 (1973).

313. Yoshio Okamoto and C. G. Overberger, *Macromolecules*, 7, 614 (1974).

314. C. G. Overberger, Thomas J. Pacansky, J. Lee, T. St. Pierre, and S. Yaroslavsky, *J. Polymer Sci., Symposia*, 46, 209 (1974).

315. C. G. Overberger and M. J. Han, *Pure and Applied Chemistry*, 39, 33 (1974).

316. C. G. Overberger, *J. Polymer Sci., Symposia*, 45, 39 (1974).

317. C. G. Overberger, Y. Okamoto, and V. Bulacovschi, *Macromolecules*, 8, 31 (1975).

318. C. G. Overberger and T. J. Pacansky, *J. Polymer Sci.*, Chemistry Edition, 13, 931 (1975).

319. C. G. Overberger and Thomas W. Smith, *Macromolecules*, 8, 401 (1975).

320. C. G. Overberger and Thomas W. Smith, *Macromolecules*, 8, 407 (1975).

321. C. G. Overberger and Thomas W. Smith, *Macromolecules*, 8, 416 (1975).

322. C. G. Overberger and Eugene Sincich, *J. Polymer Sci.*, Chemistry Edition, 13, 1783 (1975).

323. L. J. Mathias and C. G. Overberger, *Synthetic Communications*, 5(6), 461 (1975).

324. C. G. Overberger and M. J. Han, *J. Polymer Sci.*, Chemistry Edition, 13, 2251 (1975).

325. C. G. Overberger, Thomas W. Smith, and Kenneth W. Dixon, *J. Polymer Sci., Symposia*, 50, 1 (1975).

326. C. G. Overberger and M. J. Han, *J. Polymer Sci., Symposia*, 51, 155 (1975).

A. Peterlin

SOLUTION PROPERTIES OF POLYMERS

1. Rigid Spheroids

My Ph. D. thesis work was concerned with the intrinsic viscosity of a suspension of rigid spheroids with symmetry axis a_1 and perpendicular axis a_2.[1-5] The most important result was the derivation of angular distribution of symmetry axis as function of axial ratio $p = a_1/a_2$ and dimensionless velocity gradient parameter $\sigma = \dot{\gamma}/D_r$ where $\dot{\gamma}$ is the transverse velocity gradient and D_r is the rotational diffusion coefficient of the symmetry axis of the spheroid. The steady state distribution function permits the calculation of the gradient dependence of intrinsic viscosity $[\eta]$, streaming birefringence $[\Delta n]$ and extinction angle χ. The intrinsic viscosity calculated according to the then accepted rules of excess energy dissipation by the spheroids exhibited an initial increase with a maximum and subsequent drop to a finite value $[\eta]_\infty < [\eta]_0$. Much later, Kuhn[1] suggested a modification of energy dissipation by inclusions of the contribution of Brownian motion which increased $[\eta]_0$ by a factor 2 and made disappear the initial increase and maximum of intrinsic viscosity. The final formulation of excess energy dissipation by Saito[2] was used by Scheraga[3] for the correct calculation of gradient dependence of intrinsic viscosity of spheroid suspension.

The streaming birefringence exhibits saturation effects when the extinction angle χ approaches zero as the particles are as much as possible oriented in the flow direction. But the orientation is dynamic as the prolate particles possess the slowest rotational velocity in the flow direction. Hence they remain for a much

longer while in this direction than in any other direction. With
oblate particles, the preferred orientation of the symmetry axis
is in the flow plane perpendicular to the flow direction. An
applied electric field changes this distribution and hence modifies
intrinsic viscosity.[47]

The initial values $[\Delta n]$ and $\chi = 45°$ corresponding to $\sigma \to 0$
immediately apply to low molecular weight liquids.[6] As a consequence of the much oversimplified application of an isolated
particle theory to the pure liquid the parameters p and $\alpha_1 - \alpha_2$
(optical anisotropy of the molecule) are "effective" values of the
molecule which may substantially differ from those obtained in the
gas phase.

2.   Non-Gaussian Coil

The non-linear dependence of osmotic pressure of the monomer
on concentration yields in first approximation a finite second
virial coefficient A_2. Its introduction in the spatial distribution
function of monomers of a randomly coiled linear polymer yields for the
the mean square of the end-to-end distance, h^2, in addition to the
conventional linear term propostional to M a term proportional to
$M^{1/2}$.[10] Repeated application of this procedure on longer and
larger sections of the chain yields h^2 proportional to $M^{1+\varepsilon}$ with
ε proportional to A_2. With vanishing A_2, i.e., in a thermodynami-
cally ideal system, $\varepsilon = 0$. The coil is Gaussian. The better the
solvent, the higher A_2 and ε.[10,18] For such a case, a relatively
good approximation of average square of interbead distances is
obtained by $m^{1+\varepsilon}$ with $m = |j-k|$ segments between the beads j and
k. The relationship between h and gyration radius R reads $h^2 =$
$(2+\varepsilon)(3+\varepsilon)R^2$.[45] Such a model was extensively used for the calculation of intrinsic viscosity and light scattering. The coil
extension can be described by a corrected distribution function
which besides the exponential term contains as a factor a series
expansion in powers of intrabead distances.[48,49]

3.   Necklace Model of Randomly Coiled Macromolecules

Zimm's model[4] with isotropic and constant hydrodynamic interaction yields a system of partial linear differential equations
which can be separated into a system of Z independent partial
differential equations by introduction of normal coordinates. The
mathematical problem is the simultaneous diagonalization of the
matrices of hydrodynamic interaction $\underset{\sim}{H}$ and of the product $\underset{\sim}{H}\underset{\sim}{A}$ where
$\underset{\sim}{A}$ is the singular symmetric matrix of elastic forces between any
two beads. The definite positive symmetric matrix $\underset{\sim}{H}$ can be
expressed as the square of a symmetric matrix $\underset{\sim}{C}$ and the symmetric

product $\underset{\sim}{C}\underset{\sim}{A}\underset{\sim}{C}$ diagonalized by a unitary transformation $\underset{\sim}{U}$.

$$\underset{\sim}{C}\underset{\sim}{A}\underset{\sim}{C} = \underset{\sim}{U}^{-1}\underset{\sim}{D}\underset{\sim}{U} = \underset{\sim}{U}^T\underset{\sim}{D}\underset{\sim}{U}$$

The product $\underset{\sim}{Q} = \underset{\sim}{C}\underset{\sim}{U}^{-1}$ performs the desired diagonalization yielding $\underset{\sim}{A} = \underset{\sim}{M}$ in contrast to the old guess[5] that $\underset{\sim}{M}$ equals $\underset{\sim}{A}$ of Rouse's model[6] without hydrodynamic interaction. The values of intrinsic viscosity at zero gradient and zero frequency are not affected by this result. But the values at finite $\dot\gamma$ or ω are modified as soon as internal viscosity is considered.

The frequency dependence of intrinsic viscosity shows a finite value $[\eta]_\infty$ for $\omega \to \infty$.[7] In this range the imaginary part G'' of shear modulus hence increases linearly with frequency and does not remain constant as expected on the basis of Zimm's model. Such an effect can be explained[135,147,157] by introduction of internal viscosity. The ratio of internal viscosity coefficient, ϕ, and the viscous resistance of the bead, $6 \pi a_h \eta_s$, (η_s = solvent viscosity, a_h = hydrodynamic radius of the bead) as derived from experiments on PS solutions in highly viscous Aroclor turned out to be almost independent of η_s which means that ϕ is proportional to η_s.[8] Such a result contradicts the original concept of internal viscosity based on energy barriers which are independent of η_s. The consideration of actual chain displacement at any conformational change, however, shows that the segment under consideration has to be moved not only parallel to the applied force but also in perpendicular direction.[242,273] Hence the work performed during such a displacement is higher than actually considered in the force equilibrium equation for each bead. The inclusion of this excess work, proportional to η_s, is just achieved by introduction of internal viscosity.

The concept of internal viscosity also modifies the streaming birefringence in an oscillating flow field[146,158] and the acoustic birefringence.[142]

The gradient dependence of intrinsic stress tensor is more difficult to calculate. As a consequence of coil deformation the average inverse interbead distances differ from those in solution at rest. This changes the interaction tensor $\underset{\sim}{H}$ in laminar flow with transverse or longitudinal gradient. A calculation of this effect in the whole range of β is underway. The light scattering of a deformed coil without the consideration of the change of hydrodynamic interaction was calculated extensively.[95,114,115]

4.   Simplified Necklace Model

Intrinsic viscosity[12,15,25,41,42], streaming birefringence[17], light scattering of polymer solutions in flow[56,61], acoustic

birefringence[13], and sedimentation coefficient[12,15] for randomly
coiled macromolecules were calculated on the basis of a simplified
necklace model. One considers explicitly the location of chain
ends, $\vec{r}_0$ and $\vec{r}_Z$, in the laminar flow field with transverse
($\dot{\gamma} = 0$: sedimentation; $\dot{\gamma}_t$ = const.: viscosity, streaming birefringence) and longitudinal ($\dot{\gamma}_\ell$ = const.: Trouton viscosity,
birefringence) gradient. The average location of all internal
beads and the average values of interbead distances are derived
from the coordinates of the ends: $\vec{r}_j = \vec{r}_0 + j\ (\vec{r}_Z - \vec{r}_0)/Z$.

The extremely simple expressions for intrinsic viscosity at
zero gradient and sedimentation coefficient containing in the
denominator the square root of molecular weight agree very well
with experimental data of theta solutions in the whole range from
fully drained to impermeable coil. They turned out to be very
good approximations of the more complicated expressions derived
by Kirkwood-Riseman[9] and Debye-Bueche[10] and hence were used for
the calculation of intrinsic viscosity of polydisperse samples.[11]

A combination of intrinsic viscosity and birefringence data
yields the coil dimensions in flow.[84,85,88] While polyisobutylene
expands almost as a perfectly soft coil nitrocellulose molecules
expand much less.

A thorough investigation of transient effects and equilibrium
distribution of dilute polymer solutions in electrosedimentation
was performed.[62,64]

With very short chains the coil becomes a poor approximation.
The macromolecule in such a case is better represented by a
slightly bent rod. Plotted as a function of M the intrinsic
viscosity exhibits first a nearly quadratic dependence on M[14,31,176]
and the sedimentation coefficient in some cases a negative intercept
at M = 0.[20,32,49] The angular dependence of light scattering
intensity $I(\theta)$ plotted versus the square of s = $(4\pi/\lambda)\sin \theta/2$
exhibits with increasing M the complete transition from that of
the rigid rod to that of a random coil.[26,32,33] The product $s^2 I$
of small angle x-ray scattering (SAXS) as function of s exhibits
the characteristic transition from horizontal (independent of s)
to inclined (proportional to s) section at sa = 2 if the contour
length L of the molecule is much larger than the persistence length
a, i.e., L/a > 20 so that the coil is well developed.[72]

With finite gradient the experimental viscosity data show a
finite decrease and those of streaming birefringence a more than
linear increase. The effects can be caused by coil rigidity as
measured by internal viscosity[12], the finite extensibility of the
chain (saturation effects)[136], the change of hydrodynamic interaction.[50] The second cause becomes important only at extremely

high values of the dimensionless gradient parameter

$$\beta = M[\eta]\eta_s \dot{\gamma}/NkT, \text{ i.e. at } \beta \sim (L/a)^{1/2}.$$

It yields a molecular weight dependence of the effect.[73,75] The
change of hydrodynamic interaction with coil expansion in flow
yields a minimum of intrinsic viscosity with a subsequent increase
to a maximum which is higher and located at higher β the higher
M.[69,79,102,118] In spite of some experimental support of such a
theoretical prediction,[109,168,175] one has the feeling that the
calculated effect is an artifact caused by the neglection of the
anisotropy of hydrodynamic interaction which always yields a
decrease of intrinsic viscosity.

In the case of longitudinal gradient, the viscosity increases
very rapidly and goes to infinity at $\beta = .5$ for a coil with un-
limited extensibility. Finite chain length limits the increase up
to a limiting value $[\eta]_\infty/[\eta]_o \sim L/a \sim M.$[131,141] The rapid decrease
of hydrodynamic interaction with coil expansion produces an
instability of coil deformation.[292] At a critical β_{crit} the coil
discontinuously expands to an almost fully extended conformation
yielding a substantially higher Trouton viscosity. The rapid
increase of viscosity by extended chains most likely explains the
drag reduction of turbulent flow by extremely small addition of
high molecular weight polymer.[205] In laminar flow regime the
viscous resistance of the solution is practically identical with
that of the solvent because the concentration of polymer is so
small. In turbulent flow, however, the flow field in the vicinity
of an eddy lacks the rotational component. As a consequence the
long molecules get almost fully extended and hence in spite of low
concentration increase so much the local viscosity that the eddies
lose energy by frictional resistance at a substantially higher rate
than a liquid without the polymer. The ensuing reduction of turbu-
lance reduces the drag, i.e., the energy requirement which in a
polymer free liquid is used for the formation of large eddies.

5. Concentration Effects

A master curve independent of concentration exists for specific
birefringence $\Delta n/c$ and extinction angle χ if the data are plotted
versus $(\eta - \eta_s)\dot{\gamma}/c.$[34-38,195,196] Such plots lead to the establishment of the rheooptical law, i.e., the proportionality between the
excess optical and excess mechanical stress tensor of sheared
polymer solutions independent of concentration. At low concentration
the law only applies in matching solvents where the form factor of
optical anisotropy is negligible as compared with the intrinsic
optical anisotropy of the macromolecule.

In very viscous solvents the time dependence of apparent

viscosity of high molecular weight polymer solutions may show very peculiar effects caused by gradually increasing chain entanglement in shear flow.[83,86,91,116,117,123,159,160,172,184]   After an induction period the viscosity starts to rise quite appreciably as a consequence of network formation. At higher concentration the gelation continues with prolonged shearing yielding a steadily increasing viscosity. At lower concentration a maximum is reached with subsequent slow decrease to a value close to the initial viscosity. In this final stage the liquid is composed of granular elements making the free surface fairly rough. The shearing has broken the network into smaller almost spherical units which rotate in the flow and hence disturb it much less than did the network. The few macromolecules still connecting the units get so much strained that in the case of crystallizable polymer they give raise to row nucleation.[13] The network and the units disintegrate slowly in the liquid at rest so that after a while the experiment can be repeated.

SOLID STATE OF POLYMERS

1. Polymer Crystal Physics

The axial oscillation and rotation of polymer chains in the crystal lattice yield at any temperature a minimum of Gibbs free energy at a finite long period L which in linear polyethylene (PE) for a reasonable choice of surface free energy well agrees with the experimentally observed lamella thickness.[67,68,80,81,120] This minimum very likely does not play any substantial role in polymer crystallization which is mainly ruled by kinetic considerations. But it seems to be of paramount importance in the dramatic adjustment of L to the adiabatic local temperature during plastic deformation.[155,171,188,191,214,254]

The long period growth during annealing is a linear function of $\log t_A$ with the slope rapidly increasing with temperature. Such a time dependence results as a consequence of cooperative jumps of whole straight sections of the folded macromolecule.[87,101,108] The resistance to such a jump increases as L which explains the logarithmic dependence on annealing time t_A. The driving force is the reduction of surface free energy which may lead to an optimum size of crystal blocks[284] with the ratio of axial (L) and lateral (b) dimensions equal to the ratio of surface free energy of the fold containing faces (σ_e ~60-90 erg/cm^2) to that of the lateral faces (σ ~ 12 erg/cm^2).

The location and amount of amorphous component in PE single crystals was studied extensively.[181,193,206,234,256,262,267] The statistical analysis of regular and loose loops shows that a large fraction of them must have a non-adjacent **reentry if one wishes to** explain the observed thickness of amorphous layers on both sides of lamellar crystals.

NMR work on semicrystalline polymers was started with the intention to study the difference of chain mobility in the amorphous and crystalline component. The first results were the detection of the irreversible increase of amount of amorphous component and its mobility with annealing,[70] the mobilization of amorphous component by swelling liquids,[100] and the substantial difference in mobility between linear and branched PE.[66,212] The high orientation of crystal lattice of drawn samples allowed the investigation of angular dependence of NMR signal. Polymers with zig-zag chain conformation in the crystal lattice, i.e., PE,[97,106,139,148] PVF$_2$,[133] nylon[229,235] yield maxima of second moment $<\Delta H^2>$, those with helical chain conformation, i.e., POM,[106,216] PEO,[152] PP,[105] minima at 0° and 90° to the axis of orientation. The temperature dependence of the angular distribution of $<\Delta H^2>$ can be explained by a flip-flop motion of PE chains coupled with a longitudinal translation by half the period, i.e., $c/2$, but not by an oscillation or uniform rotation around the chain axis.[200,202] In the case of PEO the helical chains start to rotate about the helical axis with increasing temperature thus transforming the low temperature maxima into high temperature minima and vice versa.[152]

Laser Raman scattering on single crystals and fibrous PE yields an absorption line which corresponds to the accordion type vibration of the stretched chain sections of the crystal lattice.[218,228,278] The line is sharp in single crystals with the frequency decreasing as the inverse lamella thickness and rather broad in drawn samples as a consequence of the very uneven boundary between amorphous and crystalline domains. As expected from SAXS, annealing makes the line sharper because it smooths the boundary.

2. Plastic Deformation

Plastic deformation of unoriented lamellar samples[129,140,145,151,153,162,164,186,197,209,211,213,215,221,224,225,250,259,263,272,289] yields fibrous material.[150,165,182,243,248,270,283,290,293] The transformation mainly takes place by breaking the lamellae and pulling out of them microfibrils with alternating crystalline blocks and amorphous layers as demonstrated by iodine staining[121] and dark field electron microscopy.[162] The amorphous layers are bridged by a large fraction of taut tie molecules unfolded during breaking of lamellae into small folded chain blocks.[186] Their fraction increases with draw ratio as shown by FNA etching and subsequent DSC of melted and recrystallized debris.[149,167,177,207] Their tautness is responsible for the high axial elastic modulus which may reach 80 GPa.[14,15] Annealing[156] reduces the elastic modulus to the value of undrawn material (2 GPa). As

expected, the orientation of amorphous chain segments investigated
by IR dichroism, mainly reflecting the presence of taut tie
molecules, is very closely related with changes of elastic
modulus.[208,226,237,238] There is hardly any significant correlation with the orientation of crystalline component. The existance
of almost fully stretched taut tie molecules allows a denser packing, hence a higher density and lower heat content of amorphous
component.[122,125,138] The ensuing reduced fractional free volume
reduces sorption and diffusion of gases and vapors.[137,161,230,259]

Drawing changes the long period to a value corresponding to
the maximum adiabatic temperature the volume element attains as a
consequence of heat generation by the plastic deformational work
work.[155,171,188,191,192,214,254] This temperature is still so far
below the melting point that the sample even in the smallest
elements remains solid.[213,224] The adjustment of long period can
be only understood if one assumes that the chain shearing during
plastic deformation so much facilitates the axial motion of chains
that the blocks assume the length corresponding to thermodynamic
minimum of free energy.

Each stack of parallel lamellae is gradually transformed into
a bundle of highly aligned microfibrils densely packed into a
fibril.[164,197] Electron microscopy reveals the existance of type
II crazes, i.e., cracks bridged by the great many microfibrils[209,257]
which grow primarily in thickness and much less in width in contrast
to type I crazes in amorphous samples. Each fibril differs a little
from its neighbors in draw ratio and lateral lattice orientation
as demonstrated by electron microscopy of sample surfaces after
annealing.[151,164]

Plastic deformation of fibrous structure slightly deforms the
microfibrils by shear displacement of chains in the folded chain
crystal blocks thus reducing the sharpness of the meridional
maximum of SAXS.[171,188] It displaces axially the fibrils and
deforms them by shearing forces, i.e., by axial displacement of
microfibrils.[277,283,289,295,296,299] The former effect is much
larger than the latter because the acting forces are larger and the
resistance to deformation is less as a consequence of smaller
surface to cross-section ratio ($\sim 1/\lambda^{*3/2}$). It reduces the
structural defects caused by the ends of microfibrils which are
mainly located at the outer boundary of fibrils. Hence the fibrils
get better packed laterally. This drastically increases the
resistance to further axial displacement. The fibrous material
becomes tougher requiring steadily increasing drawing stress up to
the final fracture. The much smaller displacement of microfibrils
enormously extends the interfibrillar tie molecules so that each
passes through a great many amorphous layers. Hence their fraction
per amorphous layer may become equal or even larger than that of

intrafibrillar tie molecules. This effect is larger with short
and thick fibrils, e.g., in nylon and PET, where the almost complete transformation into fibrous structure takes place at very
small daaw ratio ($\lambda^* \sim 2.5$), and smaller with linear polyethylene
($\lambda^* \sim 9$).

Drawing of any unoriented crystalline polymer in the presence
of a swelling liquid yields very small draw ratios, $\lambda \sim 2$, and a
completely different nonfibrous material. The orientation hardly
increases. But the lamellae perpendicular to the strain are
separated by holes which enormously enhanced permeability (Knudsen
type viscous flow of gases). If such a deformed sample is dried
with fixed ends it retains its extended shape and behaves like a
hard elastomer with very large recoverable strain, up to a few
hundred percent, but much better lateral mechanical strength than
the conventional hard elastomers obtained by extrusion under
extreme force and temperature gradient.[204,16]

Deformation of unoriented crystalline samples of isotactic
polypropylene (PP) at cryogenic temperatures in contact with gases
shows a pronounced crazing, lower yield stress, and higher strain
to break as long as the gas at atmospheric pressure is close to
its boiling (N_2, A, O_2) or sublimation (CO_2) point.[264,268,282,285,297]
The effect does not occur with He at its boiling point
(4.2K). The mechanical properties in He are identical with those
in vacuum. But in other gases the modifications are quite substantial so that all experimental data on polymer mechanical
properties at cryogenic temperatures are strongly affected by the
gaseous environment. The effects are largest in smectic PP
(quenched from melt), intermediate after conversion of smectic to
monoclinic phase, smallest in slowly cooled PP with large spherulites. In the first two cases the crazes are of type I as in
amorphous material while in the last case they are of type II
usually going through the cneter of spherulite along lamellae more
or less perpendicular to the applied stress and stopping at the
outer boundary of the spherulite.

Craze initiation and propagation are connected with bursts of
acoustic emission which in very brittle solids are sufficiently
strong for detection in spite of the high noise background.[300]

3. Electron Spin Resonance and Fracture

The tensile straining of fibrous polymer ruptures a great
many tie molecules, thus producing very unstable primary end-of-
chain radicals. By chain transfer they are rapidly transformed
into the more stable secondary center-of-chain radicals which are
detectable by electron spin resonance (ESR). In nylon 6 and 66
the dependence of radical population on strain and time was

investigated.[173,194,210] The number of radicals formed is
proportional to the strained volume. Per cm length of the sample
their number is many thousand times higher than the maximum
possible number of radicals in a single fracture plane.[173] The
anisotropy of ESR spectrum proved the contention that the radicals
are formed in the amorphous regions, i.e., by rupture of taut tie
molecules.[210] Due to their smaller number each of them has to
carry a substantially higher load than a chain in the crystal
lattice. During a new loading no new radicals are formed up to
the maximum strain of former runs because up to this point the still
intact tie molecules are not yet strained to the strain to
rupture.[271]

The total number of ruptured chains is too small for seriously
affecting the mechanical properties of strained sample as measured
in subsequent runs.[189,231,236,240,252,253,260,265,271,294] Their
number shows that chain rupture mainly occurs at some isolated
areas of the sample where the magnitude of the strain and the stress
concentration on tie molecules is so much higher than in the rest
of the sample that rupture can take place. Such areas seem to be
the ends of microfibrila. Here the axial connection by tie
molecules and good lateral adhesion to adjacent microfibrils are
at their minimum. Under tensile load the material separates at
such defects so that holes are formed. Their appearance is
detectable by an increase of zero order SAXS which allows an
estimate of their number and average shape.[17] Any reduction of
the number and magnitude of such defects and hence of holes formed
upon straining improves the strength of the sample. Indeed low
and high tenacity nylon 6 fibers show upon loading a large and small
zero order SAXS, respectively.[18]

Taking into consideration the actually observed linear
dependence of logarithm of lifetime on applied load for polymer
samples,[19] the non-linear correlation of these two quantities for
a covalent chemical bond,[265] the small fraction of ruptured chains,
and the identical activation energy for sample failure and plastic
drawing[20] one concludes that the failure of a sample in creep is
primarily a consequence of mutual shear displacement of adjacent
microfibrillar elements, i.e., a frictional mode, which leads to
gradual formation of microcracks at their ends. As soon as by
axial and radical coalescence of such microcracks a critical size
crack is formed the sample fails catastrophically with the
characteristic fibrillar fracture surface. The rupture of polymer
chains is the consequence but not the primary cause of microfibrillar displacement and ensuing sample failure.[270,274,277,289,
295,296]

4. Transport Properties of Polymers

The formulation of transport properties of polymeric membranes
on the basis of thermodynamics of irreversible processes avoids the
many pitfalls of Fick's formulation in non-ideal and/or non-
homogeneous membrane.[266,281,286] The negative gradient of chemical
potential as driving force yields a dependence of the diffusive
current on the gradient pressure or compositional concentration of
the diffusant but not on the gradient of its volume concentration.
Inhomogeneous membranes may even show a maximum of concentration
and hence a flow in the direction of concentration gradient so that
adherence to Fick's law leads to a drastic local variation of D up
to infinity and negative values of D in complete contrast to
physical reality. In agreement with thermodynamics but in contrast
to first Fick's law any asymmetry of membrane composition reflected
in a finite concentration gradient within the membrane leads to
asymmetry of permeability but not to a steady state flow without
any outside driving force, i.e., without a difference in diffusant
composition or applied pressure.[223,232,239,247,255]

The semiempirical connection of diffusivity

$$D = D_o \exp(-A_1/f) \sim D_{c=o} \exp[(A_1 f_1/v_2^2 f_2^2)c] = D_{c=o} \exp(\gamma_D c)$$

and hence that of diffusive permeability

$$P = [x/(x + 1)] D = [x/(x + 1)]P_o \exp(-A_1/f)$$

with the fractional free volume of the permanent-membrane system

$$f = (f_1 + xf_2)/(1 + x)$$

describes extremely well the diffusive permeability in a wide
range of membrane swelling.[185,222,244,245] Here x is the polymer/
liquid volume ratio and the indices 1 and 2 apply to liquid and
polymer, respectively. With higher swelling of membrane the
hydraulic permeability K becomes gradually larger than the equivalent diffusion permeability PV_1/RT (V_1 = molar volume of permeant,
R = gas constant, T = absolute temperature) as a consequence of
viscous flow permeability[244,245]

$$K_v = K - PV_1/RT = (M_o/W_o \rho_2 N)(x^{-1} - x_o^{-1})$$

M_o and W_o are molecular weight and effective frictional coefficient
of polymer segment of membrane, ρ_2 is the density of dry membrane,
N. is Avogadro number, x_o^{-1} a constant, i.e., the ratio below which
no viscous flow takes place.

The Huggins-Flory formulation of equilibrium between polymer
sorbent and low molecular weight sorbate was derived by Hildebrand

under the assumption of equality of fractional free volume of both components.[21] If this restriction is dropped one obtains for the change of chemical potential of the sorbate as a consequence of mixing

$$(\mu_1 - \mu_1^0)/RT = \chi_1 v_2^2 + v_2^* + \ln (1-v_2^*) - \ln p/p_T$$

$$v_2^* = 1-v_1^* = v_2/[a + (1-a) v_2]$$

$$a = f_1/f_2$$

where v is volume fraction of each component, χ_1 is the Flory-Huggins interaction parameter, and p_T is the equilibrium pressure of the coexistent sorbate gas and liquid at the temperature of the experiment. The fractional free volume f_2 of the amorphous component of a semicrystalline polymer is increased by recoverable straining and decreased by plastic deformation. As a consequence one obtains a gradual increase of sorption, diffusion, and permeability in the former[98,280] and a drastic decrease in the latter case.[137,161,230,259]

The large change of specific volume and hence of fractional free volume of the amorphous component of deformed semicrystalline polymers is a straightforward consequence of the intimate coupling of amorphous and crystalline component. The former consists of free chain ends, chain loops, and tie molecules which are on one or at both ends fixed in the crystal lattice. This fact prevents the lateral contraction of amorphous layers sandwiched between parallel lamellae so that the Poisson ratio of such amorphous component is closer to zero than to .5 characteristic for fully amorphous rubber.[287,291]

REFERENCES

1. W. Kuhn & M. Kuhn, *Helv. Chim. Acta,* 28, 97 (1944).
2. N. Saito, *J. Phys. Soc. Japan,* 4, 85 (1949); 6, 297 (1951, *J. Polymer Sci.,* 14, 212 (1959).
3. H. A. Scheraga, *J. Chem. Phys.,* 23, 1526 (1955).
4. B. H. Zimm, *J. Chem. Phys.,* 24, 269 (1956).
5. B. H. Zimm, G. M. Roe, and L. F. Epstein, *J. Chem. Phys.,* 24, 279 (1956).
6. P. E. Rouse, Jr., *J. Chem. Phys.,* 21, 1272 (1953).
7. J. D. Ferry, *Polymer Sci.,* C15, 307 (1966).
8. J. L. Scrag and J. D. Ferry, *Macromol.,* 4, 210 (1971).
9. J. C. Kirkwood and J. Riseman, *J. Chem. Phys.,* 16, 565 (1948).
10. P. Debye and A. M. Bueche, *J. Chem. Phys.,* 16, 573 (1948).

11. H. J. Marriman and J. J. Hermans, *J. Phys. Chem.*, 65, 385 (1961).
12. W. Kuhn and M. Kuhn, *Helv. Chim. Acta*, 29, 609, 830 (1946).
13. R. R. Williamson and W. F. Busse, *J. Appl. Phys.*, 38, 4187 (1967).
14. G. Capaccio and I. M. Ward, *Polymer*, 15, 233 (1974).
15. H. Kiho, personal communication.
16. J. L. Williams, H. G. Olf, and A. Peterlin, U. S. Patent #3,839,516, Oct. 1, 1974.
17. S. N. Zhurkov, V. S. Kuksenko, and A. I. Slutsker in Fracture, ed. by P. L. Pratt, Chapman & Hall, London, 1968, p. 531.
18. B. Crist, private communication.
19. S. N. Zhurkov, *Vest. Akad. Nauk SSSR*, 11, 78 (1957).
20. G. I. Barenblatt and I. M. Kerstein, Preprints Intern. Conf. Man-Made Fibers, Kalinin, USSR, May 1974, Sec. 1, p. 28.
21. J. L. Hildebrand and R. L. Scott, Solubility of Non-Electrolytes, Rheinhold Publ., New York, 1950.

LIST OF PUBLICATIONS RELATED TO POLYMERS

1.   A. Peterlin, *Naturw.*, 26, 168 (1938).
2.   A. Peterlin, *Z. Phys.*, 111, 232-263 (1938).
3.   A. Peterlin, *Kolloid Z.*, 86, 230-241 (1939).
4.   A. Peterlin and H. A. Stuart, *Z. Phys.*, 112, 1-9 (1939).
5.   A. Peterlin and H. A. Stuart, *Z. Phys.*, 112, 129-147 (1939).
6.   A. Peterlin and H. A. Stuart, *Hand-u. Jahrbuch d. Chem. Physik*, 8/1B, (Akadem. Verlagsges. Leipzig, 1943).
7.   A. Peterlin, *Zb. Prir. Dr.*, Ljubljana, 3, 28-33, (1943).
8.   A. Peterlin and M. Samec, *Kolloid Z.*, 109, 96-99 (1944).
9.   A. Peterlin, *Diss. Acad. Sci.*, Ljubljana (III), 3, 77-130 (1947).
10. A. Peterlin, *Hommage National a P. Langevin et J. Perrin, College de France*, 70-78 (1948).
11. A. Peterlin, *I. Congr. Math. Phys. Jugosl.*, Bled, 1949, 223-227.
12. A. Peterlin, *Diss. Acad. Sci.*, Ljubljana (III A), 1, 39-76 (1950).
13. A. Peterlin, *Rec. Trav. Chim.*, 69, 14-21 (1950).
14. A. Peterlin, *J. Chim. Phys.*, 47, 669-675 (1950).
15. A. Peterlin, *J. Polym. Sci.*, 5, 473-482 (1950).
16. H. A. Stuart, *J. Polym. Sci.*, 5, 543-549 (1950).
17. H. A. Stuart, *J. Polym. Sci.*, 5, 551-563 (1950).
18. A. Peterlin, *Diss. Acad. Sci.*, Ljubljana, (III A), 1, 77-96 (1950).
19. A. Peterlin, *Diss. Acad. Sci.*, Ljubljana, (III A), 2, 29-40 (1951).
20. A. Peterlin, *J. Chimie Phys.*, 48, 13-16 (1951).

21. A. Peterlin, *Glasnik Srp. Hem. Dr., Beograd*, 215-231 (1951).
22. A. Peterlin, *Kolloid Z.*, 120, 75-83 (1951).
23. A. Peterlin, *J. Polym. Sci.*, 8, 173-185 (1952).
24. A. Peterlin, *J. Polym. Sci.*, 8, 621-632 (1952).
25. A. Peterlin, *Kunstoffe*, 42, 437-444' (1952).
26. A. Peterlin, *Nature*, 171, 259 (1953).
27. A. Peterlin, *Physik der Hochpolymeren II. 3d./5, Kap.*,
    (J. Springer-Verlag, Heidelberg) 280-332 (1953).
28. A. Peterlin, *Physik der Hochpolymeren II. 3d./6, Kap.*,
    (J. Springer-Verlag, Heidelberg) 333-354 (1953).
29. A. Peterlin, *Physik der Hochpolymeren II. 3d./11, Kap.*,
    (J. Springer-Verlag, Heidelberg) 535-568 (1953).
30. A. Peterlin and H. A. Stuart, *Physik der Hochpolymeren II.
    3d./12, Kap.*, (J. Springer-Verlag, Heidelberg) 569-617 (1953).
31. A. Peterlin, *Atti 2. Congr. Int. Mat. Plast. Torino*, 72-79
    (1950).
32. A. Peterlin, *Makromol. Chem.*, 9, 244-268 (1953).
33. A. Peterlin, *J. Polym. Sci.*, 10, 425-436 (1953).
34. A. Peterlin, *Bull. Sci. Youg.*, 1, 40-41 (1953).
35. A. Peterlin and R. Signer, *Helv. Chim. Acta*, 36, 1575-1596
    (1953).
36. A. Peterlin, *J. Polym. Sci.*, 12, 45-51 (1954).
37. A. Peterlin, *Proc. II. Int. Congr. Rheol.*, 343-349 (1953).
38. A. Peterlin and M. Copic, *Reports J. Stefan Inst.*, 1, 67-73
    (1953).
39. A. Peterlin, *Makromol. Chem.*, 13, 102 (1954).
40. A. Peterlin, *Physikertagung Innsbruck*, 75-88 (1953).
41. A. Peterlin, *Z. Naturf.*, 10a, 412-419 (1955).
42. A. Peterlin, *J. Chem. Phys.*, 23, 750-751 (1955).
43. A. Peterlin, *Ricerca Scientifica*, (Supplements) 25, 3-11 (1955).
44. A. Peterlin, *J. Colloid Science*, 10, 587-601 (1955).
45. A. Peterlin, *J. Chem. Phys.*, 23, 2464 (1955).
46. A. Peterlin, *Makromol. Chem.*, 18/19, 254-261 (1955).
47. A. Peterlin, *Diss. Acad. Sci. Ljubljana* (III A), 7, 3-17 (1956).
48. A. Peterlin, *Bull. Sci. Youg.*, 2, 98 (1956).
49. A. Peterlin, *Bull. Sci. Youg.*, 2, 98-99 (1956).
50. A. Peterlin and M. Copic, *J. Appl. Phys.*, 27, 434-438 (1956).
51. A. Peterlin, *Record Chem. Progress*, 17, 109-117 (1956).
52. A. Peterlin, *Rheology I*, Ch. 15 (Academic Press, New York
    1956) 615-652 (1956).
53. A. Peterlin and M. Copic, *Bull, Sci. Youg.*, 3, 41 (1956).
54. A. Peterlin and F. Krasovec, *Reports J. Stefan Inst.*, 3, 213-
    223 (1956).
55. A. Peterlin and V. Marinkovic, *Reports J. Stefan Inst.*, 3,
    225-231 (1956).
56. A. Peterlin, *J. Polym. Sci.*, 23, 189-198 (1957).
57. A. Peterlin, *Bull. Res. Council of Israel*, 7A, 1-8 (1957).
58. F. Krasovec, N. Vene, and A. Peterlin, *Reports J. Stefan
    Inst.*, 4, 165-173 (1957).
59. A. Peterlin, *Coll. Czechoslov. Chem. Commun.*, 22, 84-98 (1957).

60. A. Peterlin, *Reports J. Stefan Inst.*, 5, 61-69 (1958).
61. A. Peterlin, W. Heller, and M. Nakagaki, *J. Chem. Phys.*, 28, 470-476 (1958).
62. A. Peterlin and M. Ribaric, *Bull. Sci. Youg.*, 4, 101-102 (1959).
63. A. Peterlin, *Progress in Biophysics*, 9, 175-237 (1959).
64. A. Peterlin and M. Ribaric, *J. Chem. Phys.*, 31, 759-767 (1959).
65. A. Peterlin, *Makromol. Chemie*, 34, 89-119 (1959).
66. A. Peterlin, F. Krasovec, E. Pirkmajer, and I. Levstek, *Makromol. Chem.*, 37, 231-242 (1960).
67. A. Peterlin and E. W. Fischer, *Z. Phys.*, 159, 272-287 (1960).
68. A. Peterlin, *J. Appl. Phys.*, 31, 1934-1938 (1960).
69. A. Peterlin, *J. Chem. Phys.*, 33, 1799-1802 (1960).
70. A. Peterlin and E. Pirkmajer, *J. Polymer Sci.*, 46, 185-194 (1960).
71. A. Peterlin, *Ann. New York Acad. Sci.*, 89, 578-607 (1961).
72. A. Peterlin, *J. Polymer Sci.*, 47, 403-415 (1960).
73. A. Peterlin, *Makromol. Chem.*, 44-46, 338-346 (1961).
74. A. Peterlin, *J. Phys. Radium*, 22, 407-411 (1961).
75. A. Peterlin, *Polymer*, 2, 257-264 (1961).
76. A. Peterlin, *Kunststoffe*, Vol. 1/2.3 (Springer Verlage, Berlin, Gottingen, Heidelberg 1962) 94-118.
77. A. Peterlin, *Kunststoffe*, Vol. 1/5.2 (Springer Verlage, Berlin, Gottingen, Heidelberg 1962) 719-733.
78. A. Peterlin, *Kolloid-Z.*, 182, 110-115 (1962).
79. A. Peterlin, *Symp. Non-Newtonian Viscometry*, Washington 1960, ASTM 299, 115-122 (1962).
80. A. Peterlin, E. W. Fischer, Chr. Reinhold, *J. Chem. Phys.*, 37, 1403-1408 (1962).
81. A. Peterlin, E. W. Fischer, Chr. Reinhold, *J. Polymer Sci.*, 62, S59-62 (1962).
82. A. Peterlin, E. Roeckl, *J. Appl. Phys.*, 34, 102-106 (1963).
83. A. Peterlin, D. T. Turner, *Nature*, 197, 388 (1963).
84. A. Peterlin, *Kolloid-Z.*, 187, 58-59 (1963).
85. A. Peterlin, *Chimia*, 17, 65-76 (1963).
86. A. Peterlin and D. T. Turner, *J. Chem. Phys.*, 38, 2315-2316 (1963).
87. A. Peterlin, *Polymer Letters*, 1, 279-284 (1963).
88. A. Peterlin, *J. Chem. Phys.*, 39, 224-229 (1963).
89. A. Peterlin, *Intern. Conf. Electromag. Scattering, Potsdam*, 10/4/62, p. 354-375, (Pergamon Press 1963).
90. A. Peterlin, *Materials Science Research*, 1-11 (1963).
91. Mrs. S. Burow, A. Peterlin, and D. T. Turner, *Polymer Letters*, 2, 67-70 (1964).
92. P. H. Geil, H. Kiho, A. Peterlin, *Polymer Letters*, 2, 71-74 (1964).
93. Chr. Reinhold, E. W. Fischer, and A. Peterlin, *J. Appl. Phys.*, 35, 71-74 (1964).

94. A. Peterlin, *J. Appl. Phys.*, <u>35</u>, 75-81 (1964).
95. A. Peterlin and Chr. Reinhold, *J. Chem. Phys.*, <u>40</u>, 1029-1032 (1964).
96. A. Peterlin, *Polymer Letters*, <u>2</u>, 359-362 (1964).
97. A. Peterlin and H. Olf, *Polymer Letters*, <u>2</u>, 409-411 (1964).
98. H. Yasuda, V. Stannett, H. Frisch, and A. Peterlin, *Makromol. Chem.*, <u>73</u>, 180-202 (1964).
99. H. Kiho, A. Peterlin, and P. H. Geil, *J. Appl. Phys.*, <u>35</u>, 1599-1605 (1964).
100. E. W. Fischer, A. Peterlin, *Makromol. Chem.*, <u>74</u>, 1-28 (1964).
101. A. Peterlin, *Makromol. Chem.*, <u>74</u>, 107-128 (1964).
102. A. Peterlin, <u>Proc. Intern. Symp. Second Order Effects in Elasticity, Plasticity and Fluid Dynamics</u>, Haifa, April 1962, p. 776-785, Pergamon Press (1964).
103. P. Ingram and A. Peterlin, *Polymer Letters*, <u>2</u>, 739-745 (1964).
104. A. Peterlin and G. Meinel, *Polymer Letters*, <u>2</u>, 751-754 (1964).
105. A. Peterlin and H. G. Olf, *Polymer Letters*, <u>2</u>, 769-774 (1964).
106. H. G. Olf and A. Peterlin, *J. Appl. Phys.*, <u>35</u>, 3108-3114 (1964).
107. A. Peterlin and G. Meinel, *J. Appl. Phys.*, <u>35</u>, 3221-3227 (1964).
108. A. Peterlin, *Polymer*, <u>6</u>, 25-34 (1964).
109. Mrs. S. Burow, A. Peterlin, and D. T. Turner, *Polymer*, <u>6</u>, 35-47 (1965).
110. A. Peterlin, H. Kiho, and P. H. Geil, *Polymer Letters*, <u>3</u>, 151-155 (1965).
111. H. Kiho, A. Peterlin, and P. H. Geil, *Polymer Letters*, <u>3</u>, 157-160 (1965).
112. H. Kiho, A. Peterlin, and P. H. Geil, *Polymer Letters*, <u>3</u>, 257-262 (1965).
113. H. Kiho, A. Peterlin, and P. H. Geil, *Polymer Letters*, <u>3</u>, 263-270 (1965).
114. Chr. Reinhold, A. Peterlin, *Physica*, <u>31</u>, 522-540 (1965).
115. Chr. Reinhold, A. Peterlin, *J. Chem. Phys.*, <u>42</u>, 2172-2176 (1965).
116. A. Peterlin and D. T. Turner, *J. Polymer Sci.*, <u>B3</u>, 517-521 (1965).
117. A. Peterlin, C. Quan, D. T. Turner, *J. Polymer Sci.*, <u>B3</u>, 521-524 (1965).
118. A. Peterlin, <u>Proc. of Fourth Intern. Cong. on Rheology</u>, Interscience Publishers, New York (1965), 123-141.
119. A. Peterlin and J. D. Holbrook, *Kolloid-Z.*, <u>203</u>, 68-69 (1965).
120. A. Peterlin and Chr. Reinhold, *J. Polymer Sci.*, <u>A3</u>, 2801-2810 (1965).
121. A. Peterlin, P. Ingram, and H. Kiho, *Makromol. Chem.*, <u>86</u>, 294-297 (1965).
122. A. Peterlin and G. Meinel, *J. Polymer Sci.*, <u>B3</u>, 783-787 (1965).

123. A. Peterlin, D. T. Turner, and W. Philippoff, *Kolloid-Z.*, 204, 21-23 (1965).

124. A. Peterlin and Chr. Reinhold, *Kolloid-Z.*, 204, 23-28 (1965).

125. A. Peterlin and G. Meinel, *J. Appl. Phys.*, 36, 3028-3033 (1965).

126. A. Peterlin, *Makromol. Chem.*, 87, 152-165 (1965).

127. A. Peterlin and G. Meinel, *J. Polymer Sci.*, B3, 1059-1064 (1965).

128. A. Peterlin, *J. Polymer Sci.*, 3B, 1083-1087 (1965).

129. A. Peterlin, *J. Polymer Sci.*, C9, 61-89, (1965).

130. P. Ingram, H. Kiho, and A. Peterlin, *Polymer*, 7, 135-149 (1966).

131. A. Peterlin, *J. Polymer Sci.*, B4, 287-292 (1966).

132. R. L. Coalson, R. H. Marchessault, and A. Peterlin, *J. Polymer Sci.*, C13, 123-131 (1966).

133. J. B. Lando, H. G. Olf, and A. Peterlin, *J. Polymer Sci.*, A-1, 4, 941-951 (1966).

134. A. Peterlin, G. Meinel, H. G. Olf, *Polymer Sci.*, B4, 399-405 (1966).

135. A. Peterlin, *Kolloid-Z.*, 209, 181 (1966).

136. Chr. Reinhold and A. Peterlin, *J. Chem. Phys.*, 44, 4338-4341 (1966).

137. A. Peterlin, H. G. Olf, *J. Polymer Sci.*, A2, 4, 587-598 (1966).

138. A. Peterlin, G. Meinel, *Appl. Polymer Symp.*, 2, 85-100 (1966).

139. H. G. Olf, A. Peterlin, *Kolloid-Z.*, 212, 12-15 (1966).

140. A. Peterlin, K. Sakaoku, *Kolloid-Z.*, 212, 51-53 (1966).

141. A. Peterlin, *Pure and Applied Chemistry*, 12, 563-586 (1966).

142. A. Peterlin, *Polymer*, 8, 21-31 (1967).

143. A. Peterlin, J. Holbrook Elwell, *J. Materials Sci.*, 2, 1-6 (1967).

144. A. Peterlin, *J. Polymer Sci.*, C15, 337-346 (1967).

145. A. Peterlin, *J. Polymer Sci.*, C15, 427-444 (1967).

146. A. Peterlin, *J. Polymer Sci.*, B5, 113-129 (1967).

147. A. Peterlin, *J. Polymer Sci.*, A2, 5, 179-193 (1967).

148. H. Olf, A. Peterlin, *Kolloid-Z.*, 215, 97-111 (1967).

149. G. Meinel, A. Peterlin, *J. Polymer Sci.*, B, 5, 197-204 (1967).

150. A. Peterlin, *Kolloid-Z. and Z. Polymere*, 216/217, 129-136 (1967).

151. K. Sakaoku and A. Peterlin, *J. Macromol. Sci.*, (Phys.) B1, 1, 103-118 (1967).

152. H. Olf, A. Peterlin, *Makromol. Chem.*, 104, 135-141 (1967).

153. A. Peterlin, *J. Polymer Sci.*, C18, 123-132 (1967).

154. K. Sakaoku, A. Peterlin, *J. Macromol. Sci.*, (Phys.) B 1(2), 401-406 (1967).

155. R. Corneliussen, A. Peterlin, *Makromol. Chem.*, 105, 192-203 (1967).

156. G. Meinel, A. Peterlin, *J. Polymer Sci.*, B 5, 613-618 (1967).

157. A. Peterlin, Chr. Reinhold, *Trans. Soc. Rheol.*, 11:1, 15-37 (1967).

158. G. B. Thurston, A. Peterlin, *J. Chem. Phys.*, 46, 4881-4885
      (1967).
159. T. Matsuo, A. Pavan, A. Peterlin, and D. T. Turner, *J. Colloid
      & Interface Sci.*, 24, 241-251 (1967).
160. A. Pavan, T. Matsuo, A. Peterlin, and D. T. Turner, *J. Colloid
      & Interface Sci.*, 24, 273-276 (1967).
161. A. Peterlin, J. L. Williams, and V. Stannett, *J. Polymer Sci.*,
      A2, 5, 957-972 (1967).
162. K. Sakaoku, A. Peterlin, *Makromol. Chem.*, 108, 234-240 (1967).
163. P. Ingram, H. Kiho, and A. Peterlin, *J. Polymer Sci.*, C 16,
      1857-1868 (1967).
164. A. Peterlin, K. Sakaoku, *J. Appl. Phys.*, 38, 4152-4164 (1967).
165. A. Peterlin, Man Made Fibers, Ed. by H. F. Mark, S. M. Atlas,
      and E. Cernia, Interscience - John Wiley, New York, 1967;
      Chap. 8, p. 283-340.
166. A. Peterlin, Angle X-Ray Scattering, Ed. by Brumberger, Gordon
      and Breach Sci. Publ., New York, 1967, p. 145-155.
167. G. Meinel, A. Peterlin, *J. Polymer Sci.*, A2, 6, 587-605
      (1968).
168. U. Bianchi, A. Peterlin, *J. Polymer Sci.*, A2, 6, 1011-1020
      (1968).
169. K. Sakaoku, H. G. Clark, and A. Peterlin, *J. Polymer Sci.*,
      A2, 6, 1035-1040 (1968).
170. A. Peterlin, *Advances in Macromol. Chem.*, 1, 225-281 (1968).
171. A. Peterlin, R. Corneliussen, *J. Polymer Sci.*, A2, 6, 1273-
      1282 (1968).
172. A. Peterlin, *J. Lubrication Technol.*, 90, 571-579, July,
      1968.
173. D. Campbell, A. Peterlin, *J. Polymer Sci.*, B, 6, 481-485
      (1968).
174. A. Peterlin, *J. Colloid & Interface Sci.*, 27, 419-423 (1968).
175. U. Bianchi, A. Peterlin, *European Polymer J.*, 4, 515-519
      (1968).
176. A. Peterlin, *J. Polymer Sci.*, A2, 6, 1759-1772 (1968).
177. G. Meinel, A. Peterlin, and K. Sakaoku, Calorimetry of Fumic
      Nitric Acid Treated Polyethylene - Analytical Calorimetry,
      Ed. by R. S. Porter and J. F. Johnson, Plenum Press, New York
      1968, p. 15-22.
178. P. Ingram, A. Peterlin, Morphology - Encyclopedia of Polymer
      Science and Technology, Ed. by N. M. Bikales, J. Wiley,
      New York, 1968, Vol. 9, p. 204-274.
179. J. L. Williams and A. Peterlin, *Makromol. Chem.*, 120, 215-
      219 (1968).
180. D. Dolar and A. Peterlin, *J. Chem. Phys.*, 50, 3011-3015
      (1969).
181. A. Peterlin, *J. Macromol. Sci.*, (Phys.) B3, 19-31 (1969).
182. A. Peterlin, *Polymer Eng. & Sci.*, 9, 172-181 (1969).
183. A. Peterlin, *Makromol. Chem.*, 124, 136-142 (1969).
184. U. Bianchi and A. Peterlin, *Kolloid Z. & Z. Polymere*, 232,
      749-752 (1969).

185. H. Yasuda, A. Peterlin, C. K. Colton, K. A. Smith, and
     E. W. Merrill, *Makromol. Chem.*, 126, 177-186 (1969).
186. A. Peterlin, *Kolloid Z. & Z. Polymere*, 233, 857-862 (1969).
187. G. A. P. Verma and A. Peterlin, *J. Polymer Sci.*, B7, 587-590
     (1969).
188. F. J. Balta-Calleja and A. Peterlin, *J. Material Sci.*, 4,
     722-729 (1969).
189. A. Peterlin, *J. Polymer Sci.*, A2, 7, 1151-1163 (1969).
190. B. Crist and A. Peterlin, *J. Polymer Sci.*, A2, 7, 1165-1186
     (1969).
191. F. J. Balta-Calleja and A. Peterlin, *J. Polymer Sci.*, A2,
     7, 1275-1278 (1969).
192. A. Peterlin and F. J. Balta-Calleja, *J. Appl. Phys.*, 40,
     4238-4242 (1969).
193. H. G. Zachmann and A. Peterlin, *J. Macromol. Sci.*, (Phys.)
     B3, 495-517 (1969).
194. G. S. P. Verma and A. Peterlin, *Kolloid-Z. & Z. Polymere*,
     236, 111-115 (1970).
195. P. Munk and A. Peterlin, *Rheol. Acta*, 9, 288-293 (1970).
196. P. Munk and A. Peterlin, *Rheol. Acta*, 9, 294-299 (1970).
197. A. Peterlin and K. Sakaoku, Surface Morphology of Cold Drawn
     Polyethylene, Clean Surfaces, Ed. by G. Goldfinger, M. Dekker,
     Inc., New York 1970, p. 1-14.
198. P. Munk and A. Peterlin, *Trans. Soc. Rheol.*, 14, 65-75 (1970).
199. A. Peterlin and J. L. Williams, *J. Colloid & Interface Sci.*,
     32, 654-656 (1970).
200. H. G. Olf and A. Peterlin, *J. Polymer Sci.*, A2, 8, 753-770
     (1970).
201. H. G. Olf and A. Peterlin, *J. Polymer Sci.*, A2, 8, 771-789
     (1970).
202. H. G. Olf and A. Peterlin, *J. Polymer Sci.*, A2, 8, 791-797
     (1970).
203. A. Peterlin and P. Ingram, *Text. Res. J.*, 40, 345-354 (1970).
204. J. L. Williams and A. Peterlin, *Makromol. Chem.*, 135, 41-47
     (1970).
205. A. Peterlin, *Nature*, 227, 598-599 (1970).
206. A. Peterlin, *Croat. Chem. Acta*, 42, 335-349 (1970).
207. A. Peterlin and G. Meinel, Calorimetry of Drawn and Rolled
     Polyethylene of High and Low Crystallinity, Analytical
     Calorimetry, Ed. by R. S. Porter and J. F. Johnson, Plenum
     Press, New York 1970, Vol. 2, p. 9-25.
208. W. Glenz and A. Peterlin, *J. Macromol. Sci.*, B4, 473-490
     (1970).
209. F. J. Balta-Calleja and A. Peterlin, *J. Macromol. Sci.*, B4,
     519-540 (1970).
210. G. S. P. Verma and A. Peterlin, *J. Macromol. Sci.*, B4, 589-
     602 (1970).
211. G. Meinel, N. Morosoff, and A. Peterlin, *J. Polymer Sci.*, A2,
     8, 1723-1740 (1970).

212. B. Crist and A. Peterlin, *J. Macromol. Sci.*, 4B, 791-810 (1970).
213. G. Meinel and A. Peterlin, *J. Polymer Sci.*, A2, 9, 67-83 (1971).
214. A. Peterlin and F. J. Balta-Calleja, *Kolloid-Z. & Z. Polymere*, 242, 1092-1102 (1971).
215. G. Meinel and A. Peterlin, *Kolloid-Z. & Z. Polymere*, 242, 1151-1160 (1971).
216. B. Crist and A. Peterlin, *J. Polymer Sci.*, A2, 9, 557-567 (1971).
217. F. J. Balta-Calleja and A. Peterlin, *Makromol. Chem.*, 141, 91-116 (1971).
218. W. L. Peticolas, G. W. Hibbler, J. L. Lippert, A. Peterlin, and H. Olf, *Appl. Phys. Letters*, 18, 87-89 (1971).
219. A. Peterlin and G. Meinel, *Makromol. Chem.*, 142, 227-240 (1971).
220. W. Glenz, N. Morosoff, and A. Peterlin, *J. Polymer Sci.*, B9, 211-217 (1971).
221. K. Sakaoku and A. Peterlin, *J. Polymer Sci.*, A2, 9, 895-915 (1971).
222. H. Yasuda, C. E. Lamaze, and A. Peterlin, *J. Polymer Sci.*, A2, 9, 1117-1131 (1971).
223. A. Peterlin and J. L. Williams, *J. Appl. Polymer Sci.*, 15, 1493-1505 (1971).
224. A. Peterlin, *J. Material Sci.*, 6, 490-518 (1971).
225. G. Meinel and A. Peterlin, *European Polymer J.*, 7, 657-670 (1971).
226. W. Glenz and A. Peterlin, *J. Polymer Sci.*, A2, 9, 1191-1217 (1971).
227. W. Glenz, A. Peterlin, and W. Wilke, *J. Polymer Sci.*, A2, 9, 1243-1254 (1971).
228. A. Peterlin, H. G. Olf, W. L. Peticolas, G. W. Hibbler, and J. L. Lippert, *J. Polymer Sci.*, B9, 583-589 (1971).
229. H. G. Olf and A. Peterlin, *J. Polymer Sci.*, A2, 9, 1449-1469 (1971).
230. J. L. Williams and A. Peterlin, *J. Polymer Sci.*, A2, 9, 1483-1494 (1971).
231. A. Peterlin, *J. Polymer Sci.*, C32, 297-317 (1971).
232. J. L. Williams, A. Schindler, and A. Peterlin, *Makromol. Chem.*, 147, 175-184 (1971).
233. A. Peterlin, Conformations of Polymer Molecules, Polymer Science and Materials, Ed. by A. V. Tobolsky and H. Mark, J. Wiley & Sons, New York 1971, p. 46-66.
234. A. Peterlin and H. G. Zachmann, *J. Polymer Sci.*, C34, 11-17 (1971).
235. H. G. Olf and A. Peterlin, *J. Polymer Sci.*, A2, 9, 2033-2042 (1971).
236. A. Peterlin, *J. Phys. Chem.*, 75, 3921-3929 (1971).
237. W. Glenz and A. Peterlin, *Makromol. Chem.*, 150, 163-177 (1971).

238. W. Glenz and A. Peterlin, *Kolloid-Z. & Z. Polymere*, <u>247</u>, 786-794 (1971).
239. A. Peterlin, *J. Appl. Polymer Sci.*, <u>15</u>, 3127-3136 (1971).
240. A. Peterlin, *Intern. J. Fracture Mech.*, <u>7</u>, 496-499 (1971).
241. A. Peterlin and Peter Munk, Streaming Birefringence - <u>Physical Methods of Chemistry</u>, Ed. by A. Weissberger and B. Rossiter, J. Wiley & Sons, New York 1972, Vol. I, Part III, C, p. 271-384.
242. A. Peterlin, *J. Polymer Sci.*, B, <u>10</u>, 101-105 (1972).
243. A. Peterlin, *Text. Res. J.*, <u>42</u>, 20-30 (1972).
244. H. Yasuda, H. G. Olf, B. Crist, C. E. Lamaze, and A. Peterlin, Movement of Water in Homogeneous Water-Swollen Polymers - <u>Water Structure at the Water-Polymer Interface</u>, Ed. by H. H. G. Jellinek, Plenum Pub. Co., 1972, p. 39-55.
245. A. Peterlin, H. Yasuda, and H. G. Olf, *J. Appl. Polymer Sci.*, <u>16</u>, 865-870 (1972).
246. A. Peterlin, *J. Elastoplastics*, <u>4</u>, 112-130 (1972).
247. A. Peterlin, H. G. Olf, *J. Macromol. Sci.*, <u>B6</u>, 571-582 (1972).
248. A. Peterlin, *J. Elastoplastics*, <u>4</u>, 163-179 (1972).
249. N. Morosoff, K. Sakaoku, and A. Peterlin, *J. Polymer Sci.*, A-2, <u>10</u>, 1221-1236 (1972).
250. N. Morosoff and A. Peterlin, *J. Polymer Sci.*, A-2, <u>10</u>, 1237-1254 (1972).
251. K. Sakaoku and A. Peterlin, *Makromol. Chem.*, <u>157</u>, 131-139 (1972).
252. A. Peterlin, *J. Macromol. Sci.*, <u>B6</u>, 583-598 (1972).
253. A. Peterlin, *Intern. J. Fracture Mech.*, <u>8</u>, 235-236 (1972).
254. F. J. Balta-Calleja, A. Peterlin, and B. Crist, *J. Polymer Sci.*, A-2, <u>10</u>, 1749-1756 (1972).
255. A. Peterlin, *Kolloid-Z. & Z. Polymere*, <u>250</u>, 553-556 (1972).
256. W. M. Ewers, H. G. Zachmann, and A. Peterlin, *J. Macromol. Sci.*, <u>B6</u>, 695 (1972).
257. A. Peterlin, Plastic Deformation of Unoriented Crystalline Polymers Under Tensile Load - <u>Adv. Polymer Sci. & Engineering</u>, Ed. by K. D. Pae, D. R. Morrow, and Y. Chen, Plenum Press, New York-London 1972, pp. 1-20.
258. A. Peterlin, *Annual Rev. of Materials Sci.*, <u>2</u>, 349-380 (1972).
259. A. Peterlin and J. L. Williams, *Brit. Polymer J.*, <u>4</u>, 271-277 (1972).
260. A. Peterlin, Microfibrillar Structure, Radical Formation, and Fracture of Highly Drawn Crystalline Polymers - <u>Structure and Properties of Polymer Fibers</u>, Ed. by R. W. Lenz and R. S. Stein (Polymer Sci. & Technol. <u>1</u>), Plenum Press, New York-London 1972, pp. 253-265.
261. A. Peterlin, Viscosity of Polymers - <u>Physical Chemistry</u>, Ser. 1, Vol. 8, Ed. by C. E. H. Bawn, Butterworth-Univ. Park Press, London-Baltimore 1972, pp. 199-225.
262. W. M. Ewers, H. G. Zachmann, and A. Peterlin, *Kolloid-Z. & Z. Polymere*, <u>250</u>, 1187-1196 (1972).

263. K. Sakaoku, N. Morosoff, and A. Peterlin, *J. Polymer Sci.*,
      (Phys.) 11, 31-42 (1973).
264. H. G. Olf and A. Peterlin, *Polymer*, 14, 78-79 (1973).
265. A. Peterlin, ESR Investigation of Chain Rupture in Strained
      Polymer Fibers - ESR Applications to Polymer Research (Nobel
      Symp. 22), Ed. by P. O. Kinell, B. Rånby, and V. Runnstrom-
      Reio, J. Wiley & Sons (New York 1973), pp. 235-245.
266. H. Yasuda and A. Peterlin, *J. Appl. Polymer Sci.*, 17, 433-
      442 (1973).
267. A. Peterlin, *J. Colloid & Interface Sci.*, 43, 255-263 (1973).
268. H. G. Olf and A. Peterlin, *Macromolecules*, 6, 470-472 (1973).
269. A. Peterlin, *Monatshefte für Chemie*, 104, 800-813 (1973).
270. A. Peterlin, *Macromolecular Chem.*, 8, 277-304 (1973).
271. B. Crist and A. Peterlin, *Makromol. Chem.*, 171, 211 (1973).
272. A. Peterlin, *Appl. Polymer Symp.*, 20, 269-274 (1973).
273. A. Peterlin, *J. Polymer Sci.*, C43, 187-204 (1973).
274. A. Peterlin, *J. Macromol. Sci.*, B7, 705-727 (1973).
275. J. L. Williams and A. Peterlin, Membranes with Vectorized
      Flow Properties - Membranes in Separation Process, Ed. by
      C. E. Rogers and J. B. Lando, Dept. Macromol. Sci., Case
      Western Res. Univ. Cleveland, Ohio 1973, p. 14-20.
276. A. Peterlin, *Rheol. Acta*, 12, 496-502 (1973).
277. A. Peterlin, *J. Macromol. Sci.*, B8, 83-100 (1973).
278. H. G. Olf, A. Peterlin, and W. L. Peticolas, *J. Polymer Sci.*,
      (Physics) 12, 359-384 (1974).
279. P. Ingram, D. K. Woods, A. Peterlin, and J. L. Williams,
      Textile Res. J., 44, 96-106 (1974).
280. H. Yasuda and A. Peterlin, *J. Appl. Polymer Sci.*, 18, 531-546
      (1974).
281. A. Peterlin and H. Yasuda, *J. Polymer Sci.*, (Polymer Physics
      Ed.) 12, 1215-1220 (1974).
282. H. G. Olf and A. Peterlin, *J. Colloid & Interf. Sci.*, 47,
      628-635 (1974).
283. A. Peterlin, *Polymer Eng. Sci.*, 14, 627-632 (1974).
284. I. C. Sanchex, A. Peterlin, R. K. Eby, and F. L. McCrackin,
      J. Appl. Phys., 45, 4216-4219 (1974).
285. H. G. Olf and A. Peterlin, *J. Polymers Sci.*, (Physics) 12,
      2209-2252 (1974).
286. A. Peterlin, Steady State Transport Phenomena in Non-Ideal
      Permeant-Polymer Systems - Permeability of Plastic Fibers
      and Coatings, Ed. by H. Hopfenberg, Plenum Press, New York,
      1974, p. 9-34.
287. A. Peterlin, *Pure and Applied Chem.*, 39, 239-264, (1974).
288. J. L. Williams, P. Ingram, A. Peterlin, and D. K. Woods,
      Text. Res. J., 44, 370-377 (1974).
289. A. Peterlin, Mechanisms of Deformation in Polymeric Solids -
      Polymeric Materials, Ed. by E. Baer and S. V. Radcliffe,
      Amer. Soc. Metals, Metal Park, Ohio (1975), p. 175-238.
290. A. Peterlin, Molecular Aspects of Oriented Polymers - Structure
      and Properties of Oriented Polymers, Ed. by I. M. Ward, Appl.

Science Publ., London (1975), 36-55.

291. A. Peterlin, *J. Macromol. Sci.*, B11, 57-87 (1975).
292. A. Peterlin, *Makromol. Chem. Suppl.*, 1, 453-470 (1975).
293. A. Peterlin, The Composite Structure of Fibrous Material - Copolymers, Polyblends, and Composites, (Adv. in Chemistry Series No. 142), Ed. by N.A.J. Platzer, Amer. Chem. Soc., 1-13, (1975).
294. A. Peterlin, *J. Magn. Res.*, 19, 83-98 (1975).
295. A. Peterlin, *Intern. J. Fracture*, 11, 761-780 (1975).
296. A. Peterlin, *Kolloid Z. & Z. Polymere*, 253, 809-823 (1975).
297. A. Peterlin, *J. Polymer Sci.*, C50, 243-264 (1975).
298. A. Peterlin, *Annual Review of Fluid Mechanics*, 8, 35-56 (1976).
299. A. Peterlin, Crazing and Fracture of Polymer Solids - Frontiers in Materials Science, Ed. by L. E. Murr and Ch. Stein, Marcel Dekker, New York 1976, p. 491-526.
300. I. Grabec, A. Peterlin, *J. Polymer Sci.*, (Physics) 14, 651-661 (1976).
301. A. Peterlin, *Polymer Engineering & Science*, 16, 126-137 (1976).
302. J. T. Fong, A. Peterlin, *J. Research NBS*, 80 B, 273-282 (1976).

C. E. Schildknecht

Although my research for the Ph.D. at Johns Hopkins was with Frank O. Rice on photo-oxidation of acetone[1], I heard about developments in polymers from teachers E. Emmett Reid, Emil Ott, Maurice Huggins, and Neil Gordon. Beginning in September 1936 in laboratories of the Plastics Department of duPont in Arlington, N.J., I worked on the coloring of different types of plastics and later upon the reaction processes giving polymer plastics from methyl methacrylate, styrene, vinyl, chloride, and vinyl acetate as well as manufacture of polyvinyl alcohols and polyvinyl butyral. Opalescent copolymers from styrene and MMA were developed and produced[2]. Inhibition of MMA by oxygen and the oxidation products were studied[3]. These experiences, together with contacts with German polymer processes from Allied Intelligence Reports and in the General Aniline and Film Corp. (after its seizure by the U.S. in War II) gave me background for the book Vinyl and Related Polymers[4]. I also had the advantages of Herman Mark's lectures and seminars at Brooklyn Polytechnic beginning in 1941.

Stereoregulation

In G.A.F. laboratories at Easton, Pa., our evaluation of German methods for polymerizing alkyl vinyl ethers to obtain pressure sensitive adhesives led to the discovery of stereoregulation. I first disclosed in 1946 that differences in cationic polymerization conditions gave differences in properties in vinyl isobutyl and vinyl methyl ether polymers which could not be attributed to molecular weight but must result from different chemical structures, e.g., stereoisomerism[5]. Although Staudinger had not predicted stereo-

233

regulation, he had suggested in 1932 that diastereoisomerism might explain why polymers of the type --CH$_2$CHY-- did not crystallize readily. Isomerism in our vinyl isobutyl ether crystalline and amorphous polymers was mentioned briefly in an A.C.S. meeting and the resulting publication[6]. At the first Gordon Research Conference on Polymers to be held in New Hampshire (1947), in an unscheduled discussion I suggested a classification of vinyl-type polymers based upon monomer symmetry and suggested that other stereoregular, crystallizable polymers might be synthesized in the future by slow-growth ionic processes. Our later papers gave details of the flash vs. slow, proliferous processes and comparison of many properties of the normally crystalline or stereoregular and the amorphous poly-vinyl isobutyl ethers[8]. Relation of monomer structure to response in ionic and free radical systems were discussed[9].

Patents described the new slow cationic process[10] in contrast to the flash method used by Michael Otto et al. of BASF. Methods of stabilizing the vinyl ether high polymers[11], crosslinking[12], and applications in pressure sensitive adhesives[13] were developed.

The studies of vinyl ethers were part of the evaluation of "Reppe Acetylene Chemistry" by G.A.F. While alcohols react with acetylene under pressure (barring explosions) to give vinyl ethers[14] phenols with acetylene give directly resinous polymeric tackifiers for rubber of low molecular weight[15]. Another interesting product of acetylene chemistry is polyvinyl pyrrolidone which was evaluated in the laboratory, in technical sales, and in a trip to Europe in 1951 assisted by the U.S. Navy[16]. Uses of PVP were extended beyond the area of blood plasma extenders; crosslinked PVP was prepared[17]. Acrylonitrile-VP copolymers were synthesized as part of a research program, which unfortunately was soon abandoned[18,19]. Some of the copolymers had dispersion properties suggesting graft copolymers. I suggested distinguishing between interpolymers (homogeneous copolymers) and ordinary copolymer products, a practice which has been adopted frequently in U.S. patents[21].

From 1953 to 1959 I taught courses in polymer and organic chemistry at Stevens Institute of Technology, Hoboken, N.J., while consulting for Air Reduction and others. A safe plastic face mask was devised for giving anesthesia[22]. Ionic inhibition was demonstrated in the stabilization of liquid propellants such as diisopropenyl acetylene on storage[23]. Difficulties were overcome to make normally crystalline, solid homopolymers by proliferous, cationic polymerization of 2,2,2-trifluorethyl vinyl ether[24]. Conditions were devised for copolymerization of TFEVE with vinyl esters, vinyl chloride, and with acrylate esters[25]

Although I was invited to give some lectures on isomerism in vinyl ester polymers around New York City[26] and in Europe[27] before

1955, many people were sceptical. It was the surprising discovery of heat resistant, normally crystalline polyolefins[28], their helical conformation, and the prospect of a new major plastics industry which tremendously increased interest in stereoregulation. Only one partially crystalline polymer from a monoalkene of the type $CH_2=CHR$ had been reported earlier, namely that from alpha-methyl-p-methystyrene, but stereoregulation had not been attempted and it was not recognized as unusual[29]. My book in 1952 had predicted that higher melting, crystalline polymers might be prepared by special ionic polymerizations of styrenes, acrylic and other monomers of types $CH_2=CHY$ and $CH_2=CYZ$. However, I did not expect propylene to give linear polymers of high molecular weight since it is an allyl compound; this was accomplished using the Ziegler coordination or transition metal heterogeneous catalyst systems along with stereoregulation. Normally crystalline styrene, acrylic, methacrylic, and vinyl ketone polymers were soon reported using heterogeneous ionic systems[49]. Very thorough literature researches related to patent litigations on polyolefins have failed to find earlier examples of stereoregulation before our work on vinyl ether polymerizations[30].

Since 1955 researches on stereoregulation and polymer crystallinity have proliferated in the literature worldwide. I have written review articles including comparisons of stereoregulations of different vinyl ethers with those of 1-alkenes[31-35]. I spoke at numerous symposia[36], in five additional European tours[37], and at Gordon Research Conferences[38]. I have consulted on stereoregulation for major companies[39].

Some research on vinyl ether polymerization was continued largely with college undergraduates. In the slow, proliferous polymerization of vinyl isobutyl ether from liquid propane near $-70°C$ using BF_3 etherate catalyst the system is heterogeneous, but the principal seat of polymerization is a growing, homogeneous phase of swollen preformed polymer[40]. Conditions for preparing the contrasting amorphous and normally crystalline polymers were described in more detail[41]. In contrast vinyl isopropyl ether gave polymers of relatively low crystallinity even by use of slow growth polymerizations with a variety of weak catalysts[42]. Polyvinyl ethers could be cleaved to give polyvinyl alcohols which were partially crosslinked and could be characterized by iodine complexes. Aspects of polyvinyl ether morphology including spin-out of solutions in relation to structure were studied[44].

Comparison of stereoregulation in 1-alkenes with that in vinyl ethers led us to disagree with several concepts held especially in Europe. The 1-alkenes are indeed allyl compounds, which makes the coordination type catalysts the more remarkable[45,46]. These catalysts are electrophilic and the systems must be more closely re-

lated to cationic than to anionic polymerizations. Two articles
on allyl compounds were written for encyclopedias[47].

It had been assumed widely that highly stereoregular polymers
should all be normally crystalline. However, polyisobutene of regular structure only crystallizes on stretching, as do the elastomeric n-butyl vinyl ether polymers made by the slow proliferous
process. Similarly 1-pentene and 1-hexene (or butylethylene) polymers made using Ziegler catalysts are low in crystallinity and dis-
solve in hydrocarbons. Stereoregular polymers of low normal crystallinity were discussed and relations of monomer structure to
stereoregularity were reviewed[48,49]. Solution properties and hardness of contrasting highly isotactic and amorphous polyolefins and
polyvinyl ethers were studied[50].

In my books, lectures and consulting I have been much interested in tracing the history of stereoregulation. Two lectures empha-
sizing history were given in Milan and the most recent in the Centennial program of the A.C.S. Polymer Division[51].

Book Writing

Although financial returns from my books Vinyl and Related
Polymers[4] and Allyl Compounds and Their Polymers[46] have not been
great, they brought industrial consulting and fruitful contacts
with pioneers in these areas[52]. Volume 10 of the High Polymer
Series, Polymer Processes[53], sold best. The international group
of authors and efforts to describe both basic science and technology in terms as simple as possible paid off. About ten years
later when approached to write a similar book brought up to date,
the phenomenal growth of polymer science made it necessary to
largely exclude fabrication processes and to concentrate upon the
polymerization reactions. The authors for this book, Polymerization
Processes[54], were chosen in part by Irving Skeist of Skeist Laboratories, Livingston, N.J., who later found it necessary to with-
draw as coeditor. This book, now in galley proof, has been delay-
ed by difficulties in printing in the Orient. In this book effort,
as in earlier ones, my wife Althea (Schneider) Schildknecht) has
major contributions in typing and editorial assistance. She first
encountered polymers as a secretary in duPont research laboratories.

Copolymers Against Cancer

Since our research on improvement over German methods for
copolymerizing vinyl ethers with maleic anhydride in the 1940's[14],
I have experimented with maleic anhydride from time to time. By
avoiding the colored cation-radical complexes described by Takakura
we have been able to prepare in radical systems purer, colorless

maleic anhydride copolymers [55]. Copolymers made using minor proportions of divinyl ether were supplied as sodium amleate copolymers for evaluation as therapeutics to the National Cancer Institute of NIH [56]. Encouraging animal tests, especially against leu-
kemia, during the last three years have led to initiation of tests
on a larger scale. A restudy of copolymerizations of styrene with
maleic anhydride in homogeneous solution in comparison to heterogeneous growth-type copolymerizations has been undertaken in search
for examples of stereoregulation and new copolyelectrolytes for
cancer tests [57].

Telocopolymerization

I had long been puzzled by the ready reactions of alkyl vinyl
ethers with tetrachloroethylene or trichloroethylene on warming
with benzoyl peroxide giving dark, crossliked polymers. At Gettysburg we found that isopropyl or isobutyl vinyl ethers "copolymerize"
readily with $CHCl_3$ or CCl_4 in presence of radical initiators or UV
light giving viscous liquids or crosslinked solids of much higher
chlorine content than accountable by end groups (as in usual telo-
merizations). Styrene with higher proportions of halogenated hydrocarbon also telocopolymerizes; that is, the telomer acts as a telo-
gen so that the halogen compound forms links between vinyl ether or
styrene short segments. Although we reported this effect several
times [45,46,58,59] apparently it is too novel for many people to believe. A recent book on telomerization declined to include it.

Inclusion Compounds

Beginning with polyvinyl alcohols and PVP we found that more
than 20 hydrophilic polymers formed colored inclusion or clavic
compounds with $\bar{I}_3$ in aqueous solution [60-63]. Rates of formation
were studied and promotion by salts and lower temperatures was discovered. Doubt was cast upon the old theory that amylose, cellulose, and polyvinyl alcohols (PVA) require helical conformations
for the insertion of iodine. While studying PVA-iodine complexes
a novel method of making PVA fibers from foams was discovered [63].
A paper was given on applications of PVA [64], and a paper on history
of vinyl ester polymerizations with Herrmann the inventor of PVA

Other Activities

Conditions for graft copolymerization of fluoro-rubbers and PVC
with minor proportions of triallyl cyanurate with controlled crosslinking were developed for improvement of low-temperature flexibility [66]. Besides applications already mentioned, I have helped develop

for companies and government agencies adhesives, wax modifiers,
rubber tackifiers, acrylic plastics and fibers, water paints, dielectrics, medical, and dental polymers. New fibrous selectively
permeable devices were suggested[67]. Lectures were given on vinyl
monomer and polymer technology[68]. Manuscripts were criticized for
J. Polymer Science. I have also consulted for companies outside
the field of stereoregulation[69].

Besides research and teaching organic and general chemistry I
have taught polymerization courses at three colleges, served on
the A.C.S. Committee on Polymer Nomenclature since 1970, and lectured on polymer subjects as A.C.S. Visiting Scientist to Colleges
(also member of that committee). I have been chairman and organized A.C.S. symposia at Miami, Newark, Del., and Baltimore. I
have lectured on biopolymers[70], and taught organic to the remarkable David Hartman, recent M.D. from Temple Univ. I spoke on
polymer literature at a symposium at the Chemists Club, New York
City[71].

REFERENCES

1. F. O. Rice and C. E. Schildknecht, *J. Am. Chem. Soc.*, <u>60</u>, 3044
    (1938).
2. C.E.S., U.S. 2,321,048 (DuPont). Beads, bracelets, earrings,
    etc., sold by Saks Fifth Ave. et al.
3. Page 190, ref. 4.
4. C.E.S., Vinyl and Related Polymers, Wiley 1952.
5. Gordon Research Conf. on Polymers, Gibson Is., Md., July 1-5,
    1946.
6. Clyde McKinley, C.E.S., and A. O. Zoss, ACS Atlantic City,
    April 1946; Preprints Paint, *Varnish and Plastics Chem.* <u>6</u>,
    (1), 15 (1946); Ind. Eng. Chem. <u>39</u>, 180 (1947).
7. C.E.S., S. T. Gross and A. O. Zoss, *Ind. Eng. Chem.* <u>41</u>, 1998
    (1949).
8. C.E.S., A. O. Zoss, S. T. Gross, H. R. Davidson, and J. M.
    Lambert, *Ind. Eng. Chem.*, <u>40</u>, 2104 (1948).
9. C.E.S., A. O. Zoss and F. Grosser, *Ind. Eng. Chem.*, <u>41</u>, 2891
    (1949).
10. C.E.S., U.S. 2,513,820; U.S. 2,551,467; Brit. 608,202; Brit.
    611,113 (G.A.F.).
11. C.E.S., U.S. 2,395,684; U.S. 2,483,374; U.S. 2,521,950
    (G.A.F.).
12. C.E.S., U.S. 2,429,587; U.S. 2,483,374 (G.A.F.).
13. C.E.S. and I. V. Runyan, Brit. 602,424 and Brit. 640,696
    (G.A.F.).
14. C.E.S., Chapter on Vinyl Ethers in Monomers Vol. II, E. R.
    Blout, et al., ed., Interscience 1951; also ref. 6.
15. A. O. Zoss, W. E. Hanford and C.E.S., *Ind. Eng. Chem.*, <u>41</u>, 73
    (1949).

16. C.E.S., P. W. Kinney and M. L. Stecker, Periston-Type PVP, Office Tech. Services Report, PB 96,884; also in ref. 4.

17. C.E.S., U.S. 2,658,045 (G.A.F.).

18. C.E.S. and Mary L. Wallace, U.S. 2,713,573 (Celanese).

19. C.E.S., U.S. 2,768,148 (Celanese).

20. C.E.S., U.S. 2,776,947 (Celanese).

21. C.E.S., *Chem. Eng. News*, 29 (15), 1390 (1951).

22. C.E.S., Brit. 780,928 (Air Reduction); also U.S. Patent.

23. C.E.S. and R. J. Burch, U. S. 3,117,167 (Air Reduction).

24. C.E.S., U.S. 2,820,025 (Air Reduction).

25. C.E.S., U.S. 2,851,449; U.S. 2,991,278 (Air Reduction); also Ger. 1,027,872; Brit. 810,515.

26. E.g., Symposium Brooklyn Poly. Inst., Jan. 24, 1948; C.E.S. course on Vinyl-Type Polymers, Brooklyn Poly, 1949-50.

27. Univ. of Manchester, England; Plastics Inst. TNO Delft, Netherlands, 1951.

28. J. P. Hogan and R. L. Banks, Belg. 530,617 and U.S. 2,825,617 (Phillips Petroleum), cf. Chem. Week, May 14, 1955; G. Natta, P. Pino, P. Corradini, F. Danusso, E. Mantica, G. Mazzanti, and G. Moraglio, *J. Am. Chem. Soc.*, 77, 1708 (1955); cf. DuPont and Std. Oil Ind. work in ref. 46.

29. A. S. Nyquist and E. L. Kropa, U.S. 2,490,372 and Brit. 574,141 (Am. Cyan.); see ref. 4.

30. C.E.S., et al., earlier work cited in Natta et al., U.S. 3,715,344 (Montedison) and related U.S. patents.

31. C.E.S., *Chem. Eng. News*, 31, 4516 (1953).

32. C.E.S., *Ind. Eng. Chem.*, 50, 107 (1958).

33. C.E.S., *Polym. Eng. and Sci.*, 6, 240 (1966).

34. C.E.S.: Stereoregular Polymers in Encyl. of Chem., 2nd ed., G. L. Clark and G. G. Hawley, eds., Van Nostrand Reinhold 1966; 3rd ed., 1973.

35. C.E.S., Vinyl Ether Monomers and Polymers in Vol. 21, Encyl. Chem. Tech., Kirk-Othmer, 2nd ed., Wiley 1970.

36. New York, Miami, Hoboken, Philadephia, Detroit, Emeryville, Sarnia, Ottawa, Brooklyn (April 26 and Nov. 22, 1958); A.C.S. lecture tour in Michigan and Ohio 1957.

37. Including London, Nottingham, Chester, Manchester, Birmingham, Welwyn, Cardiff, Strasbourg, Mainz, Stuttgart, Cologne, Ludwigshafen, Hoeschst, Munich, Bern, Basel, Zurich, Visp, Milan.

38. Gordon Conferences in N.H. on Adhesives, Coatings and Polymers.

39. Air Reduction, American Viscose, British Nylon Spinners, Esso, Montedison, Phillips Petroleum.

40. C.E.S. and P. H. Dunn, *J. Polym. Sci.*, 20, 597 (1956).

41. C.E.S., Constance H. Lee, and W. E. Maust, Chapter in Macromolecular Synthesis, Vol. 2, J. R. Elliott, ed., Wiley, 1966.

42. C.E.S., R. H. Albright, and D. D. Oland, unpublished.

43. C.E.S., M. H. Cohen and W. E. Maust, Official Digest, *J. Paint Tech. and Eng.*, 34, (451): 800 (August 1962).

44. C.E.S., Constance Hedland Lee, Karen P. Long, and Linda C.
     Rinehart, *Polym. Eng. and Sci.*, 7 (4): 257 (1967).
45. C.E.S., Chimia (Switz.) 22 (6): 261 (1968); *Polymer Preprints*,
     12 (2): 166 (1971).
46. C.E.S., Allyl Compounds and Their Polymers (Including Poly-
     olefins), Wiley-Interscience 1973. Assistance of Mabel D.
     Reiner and Althea Schildknecht. Contribuition of 20 named
     Gettysburg College students.
47. F. Strain and C.E.S., Allyl Polymers in Encycl. Chem. Pro-
     cessing and Design Vol. 2, J. McKetta, ed., Dekker 1967;
     C.E.S. in Encycl. Chem. Tech., Kirk-Othmer Vol. 1, 3rd ed.,
     Wiley, in press.
48. C.E.S., *Polymer Preprints*, 13 (2): 1071 (1972).
49. C.E.S., *Polymer Preprints*, 13 (1): 253 (1972); Nyova Chim.
     (Italy) 48 (8): 35 (1972).
50. C.E.S., J. N. Tannebring and Robert Frederick Williams,
     Polymer Preprints, 15 (2): 421 (1974).
51. Polytechnic, Milan, June 4, 1956 and July 3, 1973; A.C.S.
     Centennial Lectures, Polymer Div., N.Y.C. April 6, 1976;
     Polymer Preprints, 17 (1): 83 (1976).
52. Among Europeans: Bunn, Evans, Fawcette, Norrish, Tuckett,
     Whinfield, Bauer, Fikentscher, Herrmann, Staudinger,
     Trommsdorff, Natta.
53. Polymer Processes, High Polymers Vol. 10, C.E.S., ed. Inter-
     science 1956.
54. C.E.S., 4 chapters in Polymerization Processes, High Polymers
     Vol. 29, C.E.S. with I. Skeist, eds.; Wiley-Interscience, in
     press.
55. C.E.S., Kathy F. Falkenstein and T. E. Stipe, 6th Middle
     Atlantic Regional ACS Meeting (MARM), Baltimore, February 1971.
56. C.E.S., Kathy F. Falkenstein and T. E. Stripe, *Preprints Org.
     Coatings and Plastics Chem.*, 33 (1): 620 (1973).
57. C.E.S., J C. S. Wood and Rosemary A. Wood, unpublished.
58. C.E.S. and W. D. Kent, 6th MARM, Baltimore, February 1971.
59. C.E.S., and K. B. Williams and W. D. Kent, *Polymer Preprints*,
     12 (2): 117 (1971); CA 78: 148280 (1973).
60. C.E.S., S. Ariemma, F. R. Litterio and R. J. Lisk, ACS New
     York City, September 1954; cf. C.E.S., *Polymer Preprints*, 1
     (2) 150 (1960); CA 57: 4833 (1962).
61. C.E.S., G. P. Colgan, C. C. DiSanto and L. R. Zegelbone, ACS
     Dallas, April 1956; cf. M. M. Zwick, *J. Appl. Polym. Sci.*, 9,
     2393 (1965).
62. C.E.S., D. B. Glenn, Jonathan Griffiths, E. W. Grotzinger,
     James Rhea, C. R. Savage and R. L. Taylor, 6th MARM Baltimore,
     February 1971.
63. C.E.S. and J. N. Tannebring, *Polymer Preprints*, 15 (2): 439
     (1974).
64. C.E.S. and G. P. Colgan, 6th MARM, Baltimore, February 1971.
65. C.E.S. and Willy O. Herrmann, 6th MARM, Baltimore, February

1971.
66. C.E.S., J. N. Tannebring and D. K. Erb, unpublished.
67. C.E.S., *Chemtech*, 6 (1): 71 (1976).
68. E.g., C.E.S., G.E. Ham, Irving Skeist, and R. H. Yocum, 6th
     MARM, Baltimore, February 1971; cf. lectures North N. J. area,
     Oil Gas J., 52 (8): 84 (1953).
69. Including Vinyl Products, Baker Chemical, Wallace and Tiernan,
     Kerr-McGee, Atlantic Refining, Conap, Nalco, Staley,
     Armstrong Cork, AMP, Resart-Ihm, U. S. Navy and Army.
70. Gettysburg College, St. Mary's College, Md., and meeting of
     Pa. Assoc. College Chem. Teachers.
71. C.E.S. in Environmental Science Technology: Information Re-
     sources, S. B. Twiner, ed., Noyes Data Corp. 1973.

J. K. Stille

This article is not intended to give a detailed account or a review of the polymer research in which I have been involved. Rather, it has been written to present a more humanistic story, as nearly as I can recall, of the people involved in the research and the development of the ideas in the polymer synthesis and the mechanisms of the polymerization reactions. The details of the conglomerate of polymer chemistry have been left in the scientific publications cited herein.

Fred Duke, a former departmental chairman at Iowa (1968-1974) once asked how my research ideas were generated; were they obtained in a flash of inspiration (light bulb, eureka, mysterious lightning, etc.), were they obtained from a dream, or were they obtained by some more gradual, obtuse process. The last is probably the most nearly correct. On occasion, I have awakened in the middle of the night with what I believed at the moment to be a fantastic research idea or a solution to a current problem. Immediately, I have written down the essence of the idea and gone back to sleep, secure with the knowledge that the valuable information would not be lost. The following morning, inspection of what I had written proved it to be gibberish in every case. What becomes apparent in this article, however, is that the research has developed by an interaction with my colleagues, my students and the scientific data, as it emerged.

Cyclopolymerization

Shortly before I started research at Iowa, George Butler had demonstrated that certain non-conjugated dienes would undergo free-radical cyclopolymerization to afford soluble materials containing

243

ring structures. At Illinois, Professor Marvel had polymerized
various non-conjugated dienes in order to obtain cyclic counter-
parts of some vinyl polymers and thereby obtain materials with
higher glass transition temperatures and structures which would not
unzip quite so readily.

When I arrived at Iowa one of the first research efforts was
aimed at obtaining copolymers of 1-olefins and a monomer which would
introduce unsaturation into the polymer by a Ziegler-Natta catalyst.
Gene vessel, who had come with me from Illinois looked at the polymerization of 2-phenylbutadiene[9] with the hopes that it would behave more like styrene than butadiene. He had set up the chemicals
and equipment necessary for work in the organometallic catalyst
area[4]. There was still some question at that time as to whether
or not cyclopolymerization was actually taking place (as opposed
to the formation of a ladder structure) and if a head-to-tail propagation was being followed. Another pioneering graduate student,
Dave Frey, undertook the polymerization of non-conjugated acetyl-
enes[7], since the cyclopolymerization would leave unsaturation in
the chain and provide functional handles for the degradation
(ozonolysis) and ultimately identification of the polymer structure
by identification of the fragments. A number of examples of cyclopolymerization and cyclocopolymerization of diolefins were added to
the growing list[14,18].

Also of interest at that time was the question of the ability
of other olefinic[14,18] and non-olefinic monomers to undergo the cyclopolymerization process. Thus, we were able to demonstrate that certain dialdehydes[18] and diepoxides[20,50,51] would undergo the cyclopolymerization reaction by a chain reaction mechanism. Bill Culbertson was the first to show that the polymerization was possible with

biallyldiepoxide. At the time I entertained thoughts of taking a
look at cyclopolymerization reactions of a variety of biscyclic sulfides, but after exploration of the polymerization reactions of
episulfides, thietanes and thiophenes[49-51,59] Joe Empen presented
convincing arguments for avoiding this area of research.*

_____

* Actually Joe's wife, Betsy, who had to live with him, presented
  more convincing arguments.

Thermally Stable Polymers

In the 1950's and 1960's, space exploration and the aerospace industry created a need for new polymer materials which could withstand severe environmental conditions, particularly high temperatures, without degradation or loss of mechanical properties. A major part of our polymer synthesis effort from the early 1960's was devoted to the synthesis of heat-stable polymers. Since it was evident that the most thermally stable organic polymers would be those which were constructed entirely of aromatic rings or aromatic rings connected by stable linkages, three different approaches were employed to achieve this. One approach was to join aromatic small molecules by a step-reaction polymerization such that the aromatic rings were connected directly or at least by a stable link such as oxygen. The formation of aromatic rings as the synthesis responsible for the polymer-forming reaction, both through the 4+2 cycloaddition reaction and polyheterocyclization reactions, were the other two approaches.

Connecting Benzene Rings. The possibility that p-benzoquinone diazide would decompose in the appropriate solvent with the loss of nitrogen to yield poly(p-phenylene oxide) was intriguing. However, complicated reaction products were obtained in non-polar solvents. Because the monomer was not appreciably soluble in non-polar solvents, Pat Cassidy carried out these reactions in tetrahydrofuran. Analysis of the polymer revealed that the polymer was not the expected poly(phenylene oxide), and by the time we had become convinced that solvents had been incorporated into the polymer, Charlie Price, who was present when these results were discussed at a symposium in Tucson for Marvel had also reached the same conclusion. Thus, an alternating copolymer of phenylene oxide and tetrahydrofuran was obtained. When other cyclic ethers were used as solvents, alternating copolymers were also formed.

A similar reaction was carried out with p-phenylenediazocarboxylate in an effort to obtain poly(p-phenylene), but polymer containing ester linkages was also obtained.*

    The possibility that hydrogen or halogens could be stripped from benzene by a glow-discharge produced by electromagnetic radiation was brought to my attention by another member of the faculty, Jack Doyle[33].† The reactions of benzene[34] and halobenzenes[35] were quite complicated, however, and although polymeric material was produced, the polymers were not obtained by the simple removal of hydrogen, hydrogen halide, or halogen from the phenyl ring.

    <u>4+2-Cycloaddition Reactions</u>.[90] The idea that the Diels-Alder addition possibly could be useful as a polymer-forming reaction was carried over from Illinois where one of Marvel's students had attempted to make a bisbutadiene in which the diene units were connected by a phenylene unit. Some work with cyclopentadiene which we had started[5,15,24,29] convinced me that polymerization <u>via</u> a Diels-Alder reaction might be possible with a biscyclopentadiene monomer.[10] Although only low molecular weight polymers were obtained, this marked the beginning of research on the synthesis of aromatic polymers by cycloaddition reactions.

    The possibility that biscyclopentadienones possibly could be suitable double dienes in Diels-Alder polymerizations was brought

---

*   This research, which represented one of my own efforts at the bench and was never published, ended in an explosion in a cold room while I was filtering a small quantity of monomer. I decided not to press my luck further, although I did return to the bench for brief periods on several occasions years later on other projects. Although it is probably true that this near disaster was due to lack of talent, I decided not to risk any graduate students on the project since I was just getting started and my group was not too large yet.

†   My opinion of inorganic chemists rose considerably when Jack described some preliminary results that had been observed with carbon tetrachloride. (I even considered asking him to join the organic division). Jack had suggested to a high school student, Fred Swift, that carbon tetrachloride might behave similarly to some of the metal tetrahalides which were known to decompose to lower valent halides with the evolution of halogen. Several months later, after having built a radio transmitter and carrying out some experiments in his home (always in the middle of the night in order to avoid hassles with the local broadcasting station) Fred returned with the results. Although he did not have the instruments to establish the structures of the products, his deductions as to the identity of the products through vapor pressure measurements were correct, and we were excited by the possibility that the reactive intermediate might be diclorocarbene.

to my attention by Frank Harris*, a graduate student who was working on another 4+2 cycloaddition polymerization (1,3-dipole) at
the time. Ernie Becker had written a review article on the synthesis and reactions of cyclopentadienones, and work at Union Carbide
by Kraimen, Wu-Chow, and Whelan showed that certain α-pyrones,
thiophene dioxides, and cyclopentadienones could indeed produce
high molecular weight polymers. Ernie was at Brooklyn Poly when
the article was written and I am still amazed that neither Becker
nor his colleagues at Poly ever made polymers with these monomers.†

A post-doctoral research associate, Harold Mukamal and two of
very able graduate students, Ruta Rakutis and Gerry Noren carried
out research on the synthesis of phenylated polyphenylenes from bis-
cyclopentadienones.[44,52,65,70,76,88] High molecular weight polymers
that contained only phenyl rings could be obtained. These polymers

were thermally stable, but by virture of the pendant phenyl groups,
were amorphous and thus soluble in common organic solvents. Marvel
and Kovacic had made poly(p-phenylene) by different reactions and,
in contrast to our polyphenylenes containing the pendant phenyl
groups, the unphenylated polyphenylenes were brown or black, crystalline, insoluble materials. A Diels-Alder synthesis of unphenylated
polyphenylenes from bispyrones revealed that the product poly-
(p-phenylene) was indeed a crystalline, insoluble, thermally stable

---

*   At that time, our oral comprehensive examination required
research proposals from the graduate students. Frank wanted to
develop this for a proposal, but I discouraged him since our departmental rules were that the proposal could not be remotely
related to the students research.

†   I even called Becker to determine whether he or any of his
former colleagues had done any work along these lines or had any
inclinations (by that time, Becker was at the University of
Massachusetts, Boston). At that time Ernie was somewhat distain-
ful of polymer chemistry; I understand now, however, that he claims
to be a polymer chemist. He even attended the Gordon Research
Conference on Polymers in 1975.

material but was yellow or light tan.*[84] [85] [93] An important lesson
learned here was that pendant phenyl groups on a polyaromatic
structure prevented crystallinity and improved solubility.

   This Diels-Alder polymer synthesis was carried out to give poly-
imides[63], ladder polyaromatic hydrocarbons[79] and polyphenylenes containing a lower degree of phenyl substitution[68,69].   During this
stage in the polymer program, the research was progressing well and
nearly everything I suggested worked out, giving the graduate students great confidence in my chemical intuition.†

   The chemistry of the 1,3-dipole cycloaddition reaction, so ele-
gantly elucidated by Professor Huisgen by the early 1960's, appeared
to be especially suited for the synthesis of aromatic polymers since
the monomers could be readily synthesized, and many of the dipolar
additions gave high yields of 5-membered heterocyclic aromatic compounds.[23]   An inspection of the literature revealed that nitrili-
mines, sydnones and nitrile oxide dipoles were especially suited.
At a Gordon Research Conference on Polymers, I learned from Charlie
Overberger that he was engaged in nitrile oxide polymerization, so
we limited our research to the bis-nitrile imine and sydnone dipoles
which afforded polypyrazoles in reactions with diethylbenzenes.[38,39,]
[41,58,60,61,71,73,90] The nitrileimine-producing monomers which
afforded the highest molecular weight polymer were tetrazoles. Such
polymers were thermally stable, but exhibited slightly less thermal
stability than some of the polymers containing 6-membered heterocyclic aromatic rings. Dihydropyrazole units could be introduced

---

*   This research was carried out by the Belgian Connection
(post-doctoral research associates, Yvan Gilliams, Henk Van Kerck-
hoven, from the laboratory of Professor Smets, as well as Jean
Braham from Professor Teyssie).

†   This also had another beneficial effect. At that time we had
started playing liar's dice at our Friday afternoon beer drinking
sessions. Because of the aura of confidence which had been built
up, I was able to lie very effectively. I even ventured to ask at
one point, as I passed the dice (and a horrendous lie) to the
graduate student on my left, whether he thought I would ever lie
to him, and if so, how he could ever again have confidence in my
ability to direct his research.

into the polymer in place of pyrazole units by the reaction of the bistetrazole with divinyl benzene (in place of diethynyl benzene). The result was a polymer with much lower thermal stability, as expected, because of the introduction of non-aromatic links.

The demonstration that the 1,3-dipole addition could be applied to the synthesis of high temperature polymers is due to the research of several devoted graduate students*, Frank Harris, Mike Bedford, Loren Gotter and Augie Chen.

Polyheterocyclization Reactions. Research in polyquinoxalines originated from information that the quinoxaline ring exhibited better thermal stability than the benzimidazole ring. The latter structure had been incorporated into aromatic polymers by Marvel with great success, but at the time my intuition was that the imidazole NH was the weak spot in oxidative degradation. Jerry Williamson synthesized our first polyquinoxalines,[22,26,28] and Fred Arnold followed with a number of other polymers,[28,35,37] which further increased our understanding of the structure-property relationships with this class of materials.[76] The polyquinoxalines were especially good high temperature adhesives, and found application as composites, films, coatings, and fibers.† (U.S. 3,661,850 and 3,734,818)   Initially we felt that is was necessary to employ

$$R-\overset{O}{\underset{}{C}}-\overset{O}{\underset{}{C}}-AR-\overset{O}{\underset{}{C}}-\overset{O}{\underset{}{C}}-R \; + \; \overset{NH_2}{\underset{NH_2}{\bigcirc}} \overset{NH_2}{\underset{NH_2}{\bigcirc}} \rightarrow \left\{ \left\{ \overset{R}{\underset{N}{\bigcirc\bigcirc}} \right\} X \left\{ \overset{N}{\underset{N}{\bigcirc\bigcirc}} \overset{R}{\underset{}{}} \right\}_{Ar} \right\}$$

X = nil, O, SO₂
AR = m-,p-C₆H₄ etc.
R = H, C₆H₅

* At this point, in an effort to increase their devotion, I began to hold research seminars on Saturday mornings at 8:00 a.m., with the theory that after seminar the students would spend the remainder of the day in the laboratory, full of inspiration and enthusiasm. As an added enticement, doughnuts and coffee were provided during the seminar. Most of my students were night people, however, and knowing that they would not possibly get up at 8:00 a.m. Saturday, took the alternative of staying up all night, spending as much of it as they could in the bars of Iowa City, and Cedar Rapids. As a result the seminars were a disaster.

† My colleages at the Du Pont Experimental Station were even kind enough to spin fiber from a sample of phenylated polyquinoxaline.

high temperatures under a vacuum to produce high molecular weight
materials, but it became evident later that because of the rapid,
complete quinoxaline-forming reaction, higher molecular weight
materials could be formed under mild conditions. Here too, the same
lesson was learned as that obtained from the polyphenylenes; phenyl
substitution on the backbone greatly improved the solubility.

The theory that a doubly stranded or ladder polymer would
exhibit greater thermal stability than the analogous bead-chain
model naturally led the research to the synthesis of ladder polyquinoxalines.[74,104]   A number of ladder structures were synthesized,
[36,42,58,55,36,102] but these polymers were not as thermally stable
as had been anticipated. Furthermore, because of their rigid
structure, they had very poor solubility and therefore, were difficult to fabricate (one example of a ladder polyquinoxaline synthesis
is shown). The lack of greatly improved thermal stability could be

the result of incomplete htereocyclization in a few of the recurring units, or the fact that linear annelation decreases the aro-
maticity (delocalization energy) per ring and thus the thermal
stability.*

---

*   The work on ladder polymer was carried out initially by graduate students Gene Mainen, and Mike Freeburger. Later the first of
a series of hard-working Japanese post-doctoral research associates
(the Japanese Connection), K. Imai and M. Kurihara continued the
work. Within a few years, post-doctoral research associates K. Imai,
Y. Imai, and H. Imai had all done research at Iowa. This led to
considerable confusion, especially since there was some overlap.
The three were soon dubbed Imai I, II, and III, respectively.

In the synthesis of the thermally stable aromatic polymers, chains with varying degrees of rigidity were obtained, and some interesting structure property relationships evolved, particularly as observed through solution viscosity and glass transition temperatures.[114]

Following this, research in aromatic heterocyclic polymers was concentrated on a polyheterocyclization reaction which afforded polyanthrazolines, polyisoanthrazolines and polyquinolines. The Friedlander synthesis had been demonstrated to afford some low molecular weight polyanthrazolines by the reaction of a diacetyl aromatic compound with 2,7-diaminoisophthaladehyde. Through the research of two post-doctoral associated, Ed. Johnson and Y. Imai[121], we learned three important things: 1) The dialdehyde would not produce high molecular weight polymer, but replacement of the aldehyde function by benzoyl groups gave monomers that afforded high molecular weight polyanthrazolines. 2) The reaction was catalyzed more effectively by acid than base. 3) Phenyl substitution on the anthrazoline ring improved the solubility (the third time this had been "discovered")*.

$$PhCO{-}C_6H_2({-}COPh)({-}NH_2)_2 \; + \; RCH_2{-}\overset{O}{\overset{\|}{C}}{-}Ar{-}\overset{O}{\overset{\|}{C}}{-}CH_2R \longrightarrow \left[\text{anthrazoline ring, } Ph, Ph, R, R, N, N, Ar \right]$$

$$R = H, C_6H_5$$

The anthrazoline ring presented too rigid a unit, however, and our attention was turned to the synthesis of the less rigid polyquinolines,[127-130] which proved to have better processability. Steve Norris† and Jim Wolfe did an outstanding job of perfecting monomer syntheses, discovering the optimum polymerization conditions (including the best catalyst, which was obtained from the reaction

---

* Actually I made this observation by venturing into the laboratory for the third time since leaving Illinois. This time there were no accidents.

† A chemical grandson from Frank Harris.

of phosphorus pentoxide with m-cresol), and characterizing the
polymers. These polymers owe their high glass transition temp-

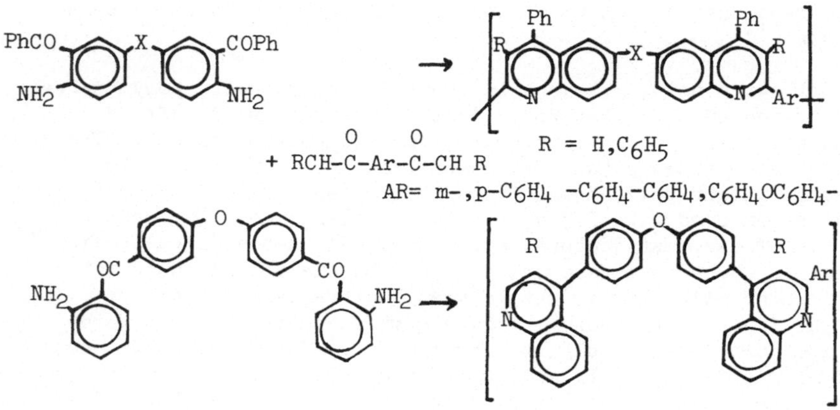

$$+ \ RCH\text{-}C\text{-}Ar\text{-}C\text{-}CH \ R$$

R = H, C_6H_5

AR= m-,p-C_6H_4 -C_6H_4-C_6H_4,$C_6H_4OC_6H_4$-

eratures (265-415°) to their rigid-chain structures, and the ability
to attain some crystallinity (Tm = 450-550°), to an ordered structure.
In contrast to the amorphous polyquinoxalines, which contain chain
disorder as a result of the alternate positions which can be taken
in the formation of the heterocyclic ring, polyquinolines can exhibit
moderately high crystallinity. These high molecular weight polymers
(Mn = 100,000) which are soluble in chloroform, N-methyl pyrrolidone,

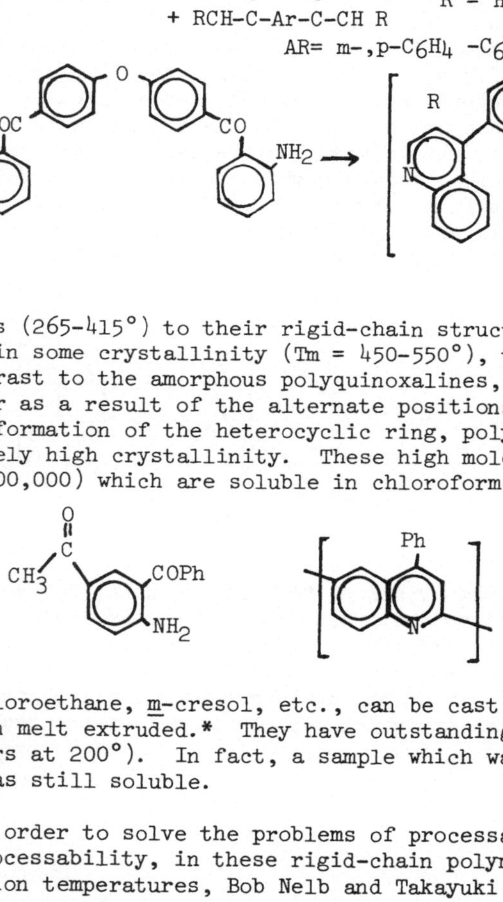

Tm = 550°

tetrachloroethane, m-cresol, etc., can be cast into films, dry spun*
and even melt extruded.* They have outstanding thermal stability
(10 years at 200°). In fact, a sample which was melt extruded at
430°C was still soluble.

    In order to solve the problems of processability, especially
melt processability, in these rigid-chain polymers that have high
transition temperatures, Bob Nelb and Takayuki Katto attempted to
develop cross-linking systems which would cure their low molecular
weight versions.[131,133] Both phenylated polyphenylenes and polyquinolines end-capped with isocyanate functions are thermosetting

_____

*   Again through the kindness of my friends at duPont.

resins by virtue of a cyclotrimetrization reaction which does not produce volatiles.[131]

$$3 \sim Ar\text{-}OCN$$

Block Polymers - Desalination Membranes

The possibility that a piezodialysis desalination membrane could be prepared from a block copolymer which, when cast into a film, would order into a charge mosaic micro-domain structure was intriguing. I'm certain that some of my interest in this area came as a result of a stimulating lecture by A. Kachalski that I heard at the Brussels IUPAC meeting, 1967. Although films of a block copolymer containing equal lengths of 2-vinylpyridine and acrylate esters formed laminar domains, hydrolysis of the ester block to acid afforded a polyampholyte which was homogeneous. At the isoelectric point, the polymer showed low flux and poor salt rejection, while on either side of the isoelectric point, salt rejection was good and flux was high.[89] A polymer, poly(N-methyl-2-vinylpyridinium-b-methacrylate) did not maintain domains, but the same type of polymer which contained additionally a styrene block did maintain domains.[101] Unfortunately, our money ran out before the work could be carried to the point where a suitable charge mosaic membrane

of the appropriate cationic and anionic domains separated by the insulating styrene matrix was realized.*

---

* Two post-doctoral research associates, Micki Kamachi and Masaru Kurihara carried out this work; Kurihara is now in charge of the desalination operation for Toray.

Charge – Transfer Polymerization[120,132]

By the late 1960's there were a number of reports in the literature of spontaneous polymerization occuring on mixing an electron donor and an electron acceptor, one or both of which was a vinyl monomer. The mechanism which was proposed for some of these polymerizations consisted of formation of a charge transfer complex, thermal one electron transfer ("T" class reaction), coupling of the radical cation and radical anion (homo-or hetero-coupling) and then anionic or cationic propatation (or both!). I was confident that we could make a contribution to the mechanism of this poly-

$$
D + A \longrightarrow \left[D \rightarrow A\right] \longrightarrow \left[D^{+}_{\cdot}\ A^{-}_{\tau}\right] \begin{array}{c} \nearrow {}^{+}D\text{-}D^{+} \\ \longrightarrow {}^{-}A\text{-}A^{-} \\ \searrow {}^{+}D\text{-}A^{-} \end{array}
$$

merization. S. Aoki from Osaka City University and R. Tarvin, a graduate student, started work at Iowa with a vacuum system while I went to Stockholm as a visiting lecturer at the Royal Institute. There, Bengt Ranby had been using flow techniques to observe short-lived radical species by esr, and from what I learned there, we were able to adapt this technique to the investigation of these polymerizations.

We were able to show[80,81,99] through esr and traping experiments, that the initiation of polymerization of vinyl ethers by 2,3-dichloro-5,6-dicyano-p-benzoquinone involved the quinone radical anion which coupled with the ether radical cation.

A similar initiation mechanism was demonstrated for the interaction of vinyl ethers with vinylidene cyanide, but in this case both an anionic and cationic homopolymerization took place in the same polymerization flask, an unusual event which was termed "cohabitory polymerization".[100] Later, Dan Chung[116,119] was able to fish some 2+2 cycloaddition products out of the polymer mixture obtained when the vinyl ethers reacted with vinylidene cyanide, thus lending support to a reaction mechanism in which the radical cation-radical anion pair could collapse to form the cyclobutane derivative or separate to initiate polymerization. The fact that a 1:1 alternating copolymer was obtained in the presence of a radical initiator supported the existence of a charge transfer complex.

At this point I speculated that a weaker donor such as a cyclic ether

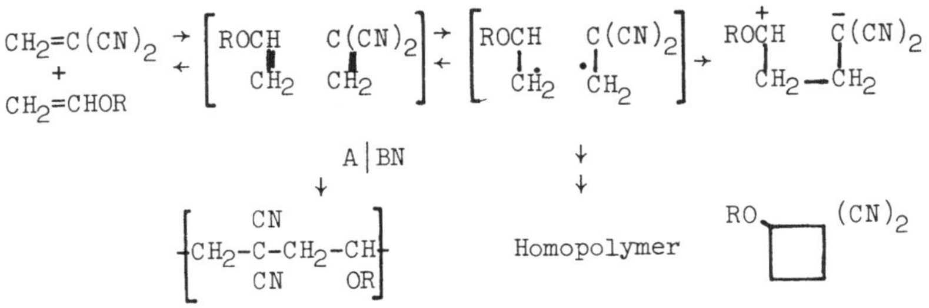

$$CH_2=C(CN)_2 + CH_2=CHOR \rightleftarrows \left[\begin{array}{cc} ROCH & C(CN)_2 \\ CH_2 & CH_2 \end{array}\right] \rightleftarrows \left[\begin{array}{cc} ROCH & C(CN)_2 \\ CH_2 & CH_2 \end{array}\right] \rightarrow \begin{array}{cc} RO\overset{+}{C}H & \overset{-}{C}(CN)_2 \\ CH_2-CH_2 \end{array}$$

$$\left[\begin{array}{c} CN \\ -CH_2-C-CH_2-CH- \\ CN \quad\quad OR \end{array}\right]$$

A | BN

Homopolymer

$$RO\underset{\square}{\quad}(CN)_2$$

might initiate the copolymerization of vinylidene cyanide. Indeed, this was the case; not only did the spontaneous anionic polymerization of vinylidene cyanade take place, but the cationic ring-opening polymerization of cyclic ehters occurred.[100,113] Drs. Kamachi, Oguni and I concluded, however, that this polymerization was not initiated by the charge-transfer mechanism.

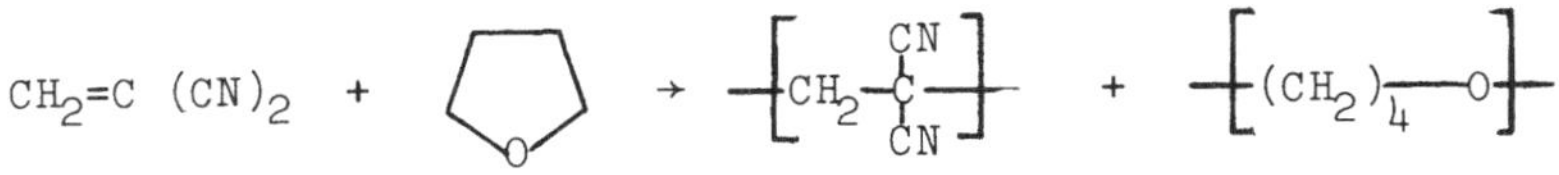

$$CH_2=C (CN)_2 + \underset{O}{\bigcirc} \rightarrow \left[\begin{array}{c} CN \\ CH_2-C- \\ CN \end{array}\right] + \left[(CH_2)_4-O\right]$$

Reactions of Polymers

During our study of 1,3-dipolar cycloaddition polymerization, and in the course of the synthesis of an AB monomer which would contain both the acetylenic and tetrazole groups, a styryl tetrazole was synthesized as a precursor to the acetylene tetrazole. It was evident that this type of monomer not only would undergo a 1,3-dipolar addition polymerization, but also the sytryl moiety would undergo typical vinyl polymerization.[72] Thus, a series of styryl tetrazoles were synthesized, and, as was anticipated, the temperature at which the tetrazole would undergo a cycloaddition with a dipolaraphile depended on the substituent group X and the position of the substituted phenyl on the tetrazole ring.

X=OCH_3, H, Cl, NO_2

The monomers would undergo radical homo- and copolymerization as well as anionic polymerization, forming living polymers. When a copolymerization was carried out with a monomer which carried a potential dipolarophile such acrylonitrile ($C\equiv N$) or butadiene (-CH=CH-), the resulting copolymer could be thermally crosslinked; the temperature at which the thermosetting reaction ensued depending on the tetrazole monomer selected.

At the present time we are engaged in research in a very exciting area of chemistry. About 40% of the publications listed at the end of this article are devoted to organic chemistry, synthetic and mechanistic. Roughly half of these "organic" publications deal with organometallic compounds[25,32,48,67,77,86,87,103,105—112,115,117,118,123—124] both with their synthesis and the mechanisms of their reactions.

Since presumably, we had become expert in two areas, polymer chemistry and organometallics, it seemed logical to join the two. The asymmetric synthesis of organic compounds in high optically yield by the reaction of a prochiral substrate with a reagent at a catalyst site which is a transition metal containing optically active ligands attached to a crosslinked polymer has a number of obvious advantages. Drs. H. Imai (Imai III), N. Takaishi and C. Bertelo have been engaged in this area of research and have had some very rewarding successes. The polymer catalyst shown will hydrogenate the prochiral substrate shown in 100% conversion to the phenylalanine precursor with 87% enantiomeric excess. The expensive catalyst is recovered by filtration and can be used again without loss of activity. The amino acid derivative is thus obtained in a pure form.[136]

Looking Back

I think it is evident from what has preceded that there were a large number of people and events that influenced the course of this research. Marvel certainly had a lot to do with the course taken by the polymer research. My students and post-doctoral research associates obviously exerted an influence and contributed much to the program in the form of ideas as well as hard work.

When I first started as an instructor at Iowa, and did much of my own research (I was much more adept then), I would start reactions in the laboratory in the morning, at 7:00 A.M. or earlier, in order to get the product out and purified by the end of the day, in spite of the teaching interruptions. On several occasions, my Department Head, Ralph Shriner, who enjoyed the early morning, would stop in the lab to visit with me. To this day, I believe that he was so impressed with my early morning industry that it greatly helped speed my first promotions. After having let him see in the laboratory early in the morning on these occasions, I was bound to keep up this facade which, at times, was very painful.*

Parallel with the polymer research, there was always a group doing research in organic syntheses and mechanisms. I felt that the organic research effort greatly aided the polymer research, because it kept me up to date in organic chemistry and allowed me to use the knowledge of organic chemistry in developing new polymerization reactions or in looking at the mechanisms of polymerization reactions. Jack Doyle was probably the one most responsible for stimulating my interest in organometallic chemistry. He had remarked that the oxymercuration reaction we had run on dicyclopentadiene in order to determine the stereochemistry[25] looked like the same reaction Chatt had run on dichlopentadiene with palladium, but the structure, including the stereoisomer obtained, had not been determined. We did determine this[32] and that was the beginning of some exciting and controversial research.

---

*   One particularly painful experience occurred after one of the annual graduate student Christmas parties at our house. The party was on Saturday night but the effects clearly reached into Monday morning.

REFERENCES

1. C. S. Marvel and J. K. Stille, *J. Org. Chem.*, 21, 1313 (1956).
2. C. S. Marvel and J. K. Stille, *J. Org. Chem.*, 22, 1451 (1957).
3. C. S. Marvel and J. K. Stille, *J. Amer. Chem. Soc.*, 80, 1740 (1958).
4. J. K. Stille, *Chem. Reviews*, 58, 541 (1958).
5. J. K. Stille and D. A. Frey, *J. Amer. Chem. Soc.*, 81, 4273 (1959).
6. J. K. Stille and E. D. Vessel, *J. Org. Chem.*, 25, 478 (1960).
7. J. K. Stille and D. A. Frey, *J. Amer. Chem. Soc.*, 83, 1697 (1961).
8. J. K. Stille and R. A. Newsom, *J. Org. Chem.*, 26, 1375 (1961).
9. J. K. Stille and E. D. Vessel, *J. Polymer Sci.*, 49, 419 (1961).
10. J. K. Stille and L. Plummer, *J. Org. Chem.*, 26, 4026 (1961).
11. J. K. Stille, *Fort. Hochpolym. Forsch.*, 3, 48 (1961).
12. J. K. Stille and T. Anyos, *J. Org. Chem.*, 27, 3352 (1962).
13. J. K. Stille, John Wiley and Sons, Inc., NY, 1962, Introduction to Polymer Chemistry.
14. J. K. Stille and D. W. Thomson, *J. Polymer Sci.*, 62, S118 (1962).
15. J. K. Stille, P. R. Kasper and D. Witherell, *J. Org. Chem.*, 28, 682 (1963).
16. J. K. Stille, P. Cassidy and L. Plummer, *J. Amer. Chem. Soc.*, 85, 1318 (1963).
17. J. K. Stille and P. Cassidy, *J. Polymer Sci.*, Part B, 1, 563 (1963).
18. J. K. Stille and R. T. Foster, *J. Org. Chem.*, 28, 2703 (1963).
19. J. K. Stille and R. T. Foster, *J. Org. Chem.*, 28, 2708 (1963).
20. J. K. Stille and B. M. Culbertson, *J. Polymer Sci.*, Part A, 2, 405 (1964).
21. J. K. Stille and R. Ertz, *J. Amer. Chem. Soc.*, 86, 661 (1964).
22. J. K. Stille and J. R. Williamson, *J. Polymer Sci.*, Part B, 2, 209 (1964).
23. J. K. Stille and T. Anyos, *J. Polymer Sci.*, Part A, 2, 1487 (1964).
24. J. K. Stille and D. R. Witherell, *J. Amer. Chem. Soc.*, 86, 2188 (1964).
25. J. K. Stille and S. C. Stinson, *Tetrahedron*, 20, 1387 (1964).
26. J. K. Stille and J. R. Williamson, *J. Polymer Sci.*, Part A, 2, 3867 (1964).
27. J. K. Stille and D. D. Whitehurst, *J. Amer. Chem. Soc.*, 86, 4871 (1964).
28. J. K. Stille, J. R. Williamson and F. E. Arnold, *J. Polymer Sci.*, Part A, 3, 1013 (1965).
29. J. K. Stille, R. J. Sevenich and D. D. Whitehurst, *J. Org. Chem.*, 30, 938 (1965).
30. J. K. Stille and C. N. Wu, *J. Org. Chem.*, 30, 1222 (1965).
31. J. K. Stille and R. A. Morgan, *J. Polymer Sci.*, Part A, 3, 2397 (1965).

32. J. K. Stille, R. A. Morgan, D. D. Whitehurst, and J. R. Doyle
     J. Amer. Chem. Soc., 87, 3282 (1965).
33. Fred Swift, Jr., R. L. Sung, J. R. Doyle and J. K. Stille,
     J. Org. Chem., 30, 3114 (1965).
34. J. K. Stille, R. L. Sung, and J. Vander Kooi, J. Org. Chem.,
     30, 3116 (1965).
35. J. K. Stille and F. E. Arnold, J. Polymer Sci., Part A, 3,
     4284 (1965).
36. J. K. Stille and E. L. Mainen, J. Polymer Sci., Part B, 4,
     39 (1966).
37. J. K. Stille and F. E. Arnold, J. Polymer Sci., Part A, 4,
     551 (1966).
38. J. K. Stille and M. A. Bedford, J. Polymer Sci., Part B, 4,
     329 (1966).
39. J. K. Stille and F. W. Harris, J. Polymer Sci., Part B, 4,
     333 (1966).
40. J. K. Stille and C. E. Rix, J. Org. Chem., 31, 1591 (1966).
41. J. K. Stille, F. W. Harris, and M. A. Bedford, J. Heterocyclic
     Chem., 3, 155 (1966).
42. J. K. Stille and E. L. Mainen, J. Polymer Sci., Part B, 4,
     665 (1966).
43. J. K. Stille and F. M. Sonnenberg, Tetrahedron Letters, 38,
     4587 (1966).
44. J. K. Stille, F. W. Harris, R. O. Rakutis, and H. Mukamal,
     J. Polymer Sci., Part B, 4, 791 (1966).
45. F. M. Sonnenberg and J. K. Stille, J. Org. Chem., 31, 3441
     (1966).
46. J. K. Stille and F. M. Sonnenberg, J. Amer. Chem. Soc., 88,
     4915 (1966).
47. J. K. Stille, F. M. Sonnenberg, and T. H. Kinstle, J. Amer.
     Chem. Soc., 88, 4922 (1966).
48. J. K. Stille and R. A. Morgan, J. Amer. Chem. Soc., 88, 5135
     (1966).
49. J. K. Stille and J. A. Empen, J. Polymer Sci., A-1, 5, 273
     (1967).
50. J. K. Stille and J. J. Hillman, J. Polymer Sci., A-1, 5, 2055
     (1967).
51. J. K. Stille and J. J. Hillman, J. Polymer Sci., A-1, 5, 2067
     (1967).
52. H. Mukamal, F. W. Harris, and J. K. Stille, J. Polymer Sci.,
     A-1, 5, 2721 (1967).
53. J. K. Stille and M. E. Freeburger, J. Polymer Sci., B, 5, 989
     (1967).
54. M. E. Freeburger, R. E. Buckles, and J. K. Stille, Tetrahedron
     Letters, 4, 431 (1968).
55. J. K. Stille and E. L. Mainen, Macromolecules, 1, 36 (1968).
56. J. K. Stille and M. E. Freeburger, J. Polymer Sci., A-1, 6,
     161 (1968).
57. J. K. Stille, Industrial Organic Chemistry, Prentice-Hall, Inc.,
     Englewood Cliffs, NJ, (1968).

58. J. K. Stille and J. A. Empen, The Chemistry of Sulfides,
    A. V. Tobolsky, ed., Interscience Publishers, NY (1968).
59. J. K. Stille and L. D. Gotter, J. Polymer Sci., B, 6, 11 (1968).
60. J. K. Stille and F. W. Harris, J. Polymer Sci., A-1, 6, 2317
    (1968).
61. J. K. Stille and M. A. Bedford, J. Polymer Sci., A-1, 6, 2331
    (1968).
62. J. K. Stille, R. O. Rakutis, H. Mukamal, and F. W. Harris,
    Macromolecules, 1, 431 (1968).
63. F. W. Harris and J. K. Stille, Macromolecules, 1, 463 (1968).
64. J. K. Stille, J. M. Unglaube and M. E. Freeburger, J. Amer.
    Chem. Soc., 90, 7076 (1968).
65. J. K. Stille and D. B. Fox, Inorg. Nucl. Chem. Letters, 5,
    157 (1969).
66. J. A. Empen and J. K. Stille, Macromolecular Synthesis, 3, 53
    (1969).
67. J. A. Empen and J. K. Stille, Macromolecular Synthesis, 3,
    56 (1969).
68. C. L. Schilling, Jr., J. A. Reed, and J. K. Stille, Macro-
    molecules, 2, 85 (1969).
69. Joe A. Reed, Curtis L. Schilling, Jr., R. F. Tarvin,
    T. A. Rettig, and J. K. Stille, J. Org. Chem., 34, 2188 (1969).
70. J. K. Stille and G. Noren, J. Polymer Sci., B, 7, 525, (1969).
71. J. K. Stille and L. D. Gotter, Macromolecules, 2, 465 (1969).
72. J. K. Stille and L. D. Gotter, Macromolecules, 2, 468 (1969).
73. J. K. Stille and L. D. Gotter, J. Polymer Sci., A-1, 7, 2493
    (1969).
74. J. K. Stille, J. Macromol. Sci.-Chem., A3(6), 1043 (1969).
75. J. K. Stille, F. W. Harris, H. Mukamal, R. O. Rakutis,
    C. L. Schilling, G. K. Noren, and J. A. Reed, Advances in
    Chemistry Series, 91, 628 (1969).
76. J. K. Stille, Encyclopedia of Polymer Science and Technology,
    11, 389 (1969).
77. J. K. Stille and D. B. Fox, J. Amer. Chem. Soc., 92, 1274
    (1970).
78. J. K. Stille and L. F. Hines, J. Amer. Chem. Soc., 92, 1798
    (1970).
79. J. K. Stille, G. K. Noren, and L. Green, J. Polymer Sci., A-1,
    8, 2245 (1970).
80. S. Aoki, R. F. Tarvin, and J. K. Stille, Macromolecules, 3,
    472 (1970).
81. S. Aoki and J. K. Stille, Macromolecules, 3, 473 (1970).
82. J. K. Stille and R. D. Hughes, J. Org. Chem., 36, 340 (1971).
83. J. L. Eichelberger and J. K. Stille, J. Org. Chem., 36, 1840
    (1971).
84. J. K. Stille and Y. Gilliams, Macromolecules, 4, 515 (1971).
85. G. K. Noren and J. K. Stille, J. Polymer Sci., D, 5, 385 (1971).
86. R. W. Fries and J. K. Stille, Syn. Inorg. Metal-Org. Chem.,
    1(4), 295 (1971).

87. L. Hines and J. K. Stille, *J. Amer. Chem. Soc.*, 94, 485 (1972).
88. J. K. Stille and G. K. Noren, *Macromolecules*, 5, 49 (1972).
89. M. Kamachi, M. Kurihara and J. K. Stille, *Macromolecules*, 5, 161 (1972).
90. J. K. Stille, *Macromol. Chem.*, 154, 49 (1972).
91. J. K. Stille, J. L. Eichelberger, J. Higgins, and M. E. Freeburger, *J. Amer. Chem. Soc.*, 94, 4761 (1972).
92. J. K. Stille and A. T. Chen, *Macromolecules*, 5, 377 (1972).
93. H. F. VanKerckhoven, Y. K. Gilliams, and J. K. Stille, *Macromolecules*, 5, 541 (1972).
94. E. W. Koos, J. P. Vander Kooi, E. E. Green, and J. K. Stille, *J. Chem. Soc., Chem. Commun.*, 1085 (1972).
95. J. K. Stille, M. E. Freeburger, W. B. Alston, and E. L. Mainen, Ch. 10 in Condensation Monomers, J. K. Stille and T. W. Campbell, eds., Wiley-Interscience, NY (1972).
96. J. K. Stille, W. A. Feld, and M. E. Freeburger, *J. Amer. Chem. Soc.*, 94, 8485 (1972).
97. J. K. Stille and K. N. Sannes, *J. Amer. Chem. Soc.*, 94, 8489 (1972).
98. J. K. Stille and K. N. Sannes, *J. Amer. Chem. Soc.*, 94, 8494 (1972).
99. R. F. Travin, S. Aoki, and J. K. Stille, *Macromolecules*, 5, 663 (1972).
100. N. Oguni, M. Kamachi, and J. K. Stille, *Macromolecules*, 6, 146 (1972).
101. M. Kurihara, M. Kamachi, and J. K. Stille, *J. Polymer Sci., Polymer Chem. Ed.*, 11, 587 (1972).
102. K. Imai, M. Kurihara, L. Mathias, J. Wittmann, W. B. Alston, and J. K. Stille, *Macromolecules*, 6, 158 (1972).
103. J. K. Stille, D. E. James, and L. F. Hines, *J. Amer. Chem. Soc.*, 95, 5062 (1973).
104. J. K. Stille, *Macromol. Chem.*, 8, 373 (1973).
105. J. K. Stille and M. T. Regan, *J. Amer. Chem. Soc.*, 96, 1508 (1974).
106. J. K. Stille and R. W. Fries, *J. Amer. Chem. Soc.*, 96, 1514 (1974).
107. J. K. Stille, F. Huang, and M. T. Regan, *J. Amer. Chem. Soc.*, 96, 1518 (1974).
108. P. K. Wong and J. K. Stille, *J. Organometal. Chem.*, 70, 121 (1974).
109. K.S.Y. Lau, R. W. Fries, and J. K. Stille, *J. Amer. Chem. Soc.*, 96, 4983 (1974).
110. P. K. Wong, K.S.Y. Lau, and J. K. Stille, *J. Amer. Chem. Soc.*, 96, 5956, (1974).
111. J. K. Stille, L. F. Hines, R. W. Fries, P. K. Wong, D. E. James, and K.S.Y. Lau, *Advances Chem. Ser.*, 132, 90 (1974).
112. J. K. Stille, M. T. Regan, R. W. Fries, F. Huang, and T. McCarley, *Advances Chem. Ser.*, 132, 181 (1974).

113. N. Oguni, M. Kamachi, and J. K. Stille, *Macromolecules*, 7, 435 (1974).

114. J. K. Stille, Proc. Symp. Macromole., Elsevier, Amsterdam, 95 (1975).

115. J. K. Stille and P. K. Wong, *J. Org. Chem.*, 40, 335 (1975).

116. J. K. Stille and D. C. Chung, *Macromolecules*, 8, 83 (1975).

117. J. K. Stille and D. E. James, *J. Amer. Chem. Soc.*, 97, 674 (1975).

118. J. K. Stille and P. K. Wong, *J. Org. Chem.*, 40, 532 (1975).

119. J. K. Stille and D. C. Chung, *Macromolecules*, 8, 114 (1975).

120. J. K. Stille, N. Oguni, D. C. Chung, R. F. Tarvin, S. Aoki, and M. Kamachi, *J. Macromol. Sci.-Chem.*, A9(5), 745 (1975).

121. Y. Imai, E. F. Johnson, T. Katto, M. Kurihara, and J. K. Stille, *J. Poly. Sci., Polymer Chem. Ed.*, 13, 2233 (1975).

122. E. W. Kuemmerle, T. A. Rettig, and J. K. Stille, *J. Org. Chem.*, 40, 3665 (1975).

123. J. K. Stille and D. E. James, *J. Organometal. Chem.*, 108, 401 (1976).

124. D. E. James, L. F. Hines, and J. K. Stille, *J. Amer. Chem. Soc.*, 98, 1806 (1976).

125. D. E. James and J. K. Stille, *J. Amer. Chem. Soc.*, 98, 1810 (1976).

126. D. E. James and J. K. Stille, *J. Org. Chem.*, 41, 1504 (1976).

127. R. G. Nelb, II, and J. K. Stille, *J. Amer. Chem. Soc.*, 98, 2834 (1976).

128. J. F. Wolfe and J. K. Stille, *Macromolecules*, 9, 489 (1976).

129. S. O. Norris and J. K. Stille, *Macromolecules*, 9, 496 (1976).

130. W. Wrasidlo and J. K. Stille, *Macromolecules*, 9, 505 (1976).

131. W. Wrasidlo, S. O. Norris, J. F. Wolfe, T. Katto, and J. K. Stille, *Macromolecules*, 9, 512 (1976).

132. J. K. Stille, R. G. Nelb, II, and S. O. Norris, *Macromolecules*, 9, 516 (1976).

133. J. K. Stille, N. Oguni, D. C. Chung, R. F. Tarvin, S. Aoki, and M. Kamachi, Ionic Polym.: Unsolved Probl. Jpn-U.S. Semin. Polym. Synth., 1st 1974, J. Furukawa and O. Vogl, eds., NY, 105 (1976).

134. K. Iwata and J. K. Stille, *J. Poly. Sci., Poly. Chem. Ed.*, 14, 2841 (1976).

135. J. K. Stille, T. A. Rettig, and E. W. Kuemmerle, *J. Org. Chem.*, 41, 2950 (1976).

136. Naotake Takaishi, Hirosuke Imai, Christopher A. Bertelo, and J. K. Stille, *J. Amer. Chem. Soc.*, 98, 5400 (1976).

137. K.S.Y. Lau, P. K. Wong, and J. K. Stille, *J. Amer. Chem. Soc.*, 98, 5832 (1976).

138. J. K. Stille and K.S.Y. Lau, *J. Amer. Chem. Soc.*, 98, 5841 (1976).

139. J. K. Stille and D. E. James, Ch. 12 in The Chemistry of Functional Groups, Supplement A: Double Bond Functional Groups, S. Patai, ed., John Wiley & Sons, Inc., London 1977.

140. J. K. Stille and A. B. Cowell, *J. Organometal. Chem.*, <u>124</u>, 253 (1977).

141. J. K. Stille and H. Mukamal, <u>Macromolecular Synthesis</u>, Vol. 6, James A. Moore, ed., John Wiley & Sons, Inc., NY, 1977, p. 45.

M. Szwarc

Studies of bond dissociation energies of organic polyatomic molecules led me to conclude that the weakest bond in toluene is the $PhCH_2$-H. Indeed, pyrolysis of toluene carried out in a flow system at pressures of 5-10 torrs resulted in its dissociation into a thermally stable benzyl radical and a hydrogen atom[1]. Thus, dibenzyl and H_2 were the main products of the decomposition.

Having established by the pyrolytic method[2] the strength of the $PhCH_2$-H bond, I turned to the question of how substituents affect the C-H bond dissociation energy of substituted toluenes. To answer this question we investigated the pyrolysis of the xylenes[1], fluorotoluenes[3], picolines[4], etc. In each case the formation of substituted dibenzyls confirmed the anticipated pattern of the decomposition initiated by rupture of the relevant C-H bond. However, the pyrolysis of p-xylene led to an unexpected result. The trap in which I hoped to collect the dimethylated dibenzyl became coated by a tough and coherent film which could be removed intact after completion of the pyrolysis.

The appearance of this film aroused my interest in polymer chemistry. In a short time I proved that the film is built from polymeric chains having the -CH_2-⟨O⟩-CH_2- moieties as recurring units. Moreover, I conclusively demonstrated that the pyrolysis of p-xylene, performed under the conditions described in ref. 1, yields p-xylelene, CH_2=⟨◯⟩=CH_2, a quinonoide hydrocarbon anticipated by theoreticians but never reported before. The proof for

265

the formation of p-xylelene[1,5] was simple. The vapors emerging
from the pyrolysis zone were carried through a glass tube, up to
3 m long, into a trap cooled to 0°C. The tube was heated electri-
cally to about 100°C and remained clear--no deposit or film was
formed on its wall. The trap contained iodine, and after completion
of the experiment p,p'-di-iodo-p-xylelene, $CH_2I.C_6H_4.CH_2I$, was found
in it as the main product. These results imply that the pyrolysis
yielded the gaseous p-xylelene which passed through the tube into
the trap where it reacted with I_2 at 0°C giving the di-iodide,

$$CH_2=\!\!\left\langle\!\bigcirc\!\right\rangle\!\!=\!CH_2 + I_2 \rightarrow CH_2I\text{-}\!\left\langle\!\bigcirc\!\right\rangle\text{-}CH_2I.$$

The facile iodine addition suggests that the quinonoide hydrocarbon

may be easily converted into a di-radical, $\cdot CH_2\text{-}\!\left\langle\!\bigcirc\!\right\rangle\text{-}CH_2\cdot$, the

latter representing the triplet state of the quinonoid singlet.
Indeed, the quantum-mechanical calculations[6] showed that the
singlet-triplet separation in this species is low.

The second evidence confirming the structure of the pyrolysis
product was furnished shortly after publication of the above results.
The pyrolysis of p-xylene yields, in addition to the poly-p-xylelene,
some minor products. From their mixture Brown and Farthing[7] isolated
an interesting hydrocarbon and identified this product by x-ray
analysis as paracyclophane,

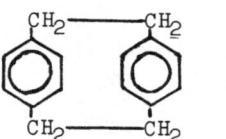

,

at that time an unknown cyclic dimer of p-xylelene. Since
p-xylelene had to be a precursor of paracyclophane, the isolation
of the latter confirms the formation of the former. The isolation
and identification of paracyclophane preceded its first conventional
synthesis by Cram[8].

The details of the mechanism of formation of p-xylelene from
p-xylene are still unknown. Undoubtedly the rupture of the C-H
bond yields the p-xylyl radical,

$$\text{p-xylene} \rightarrow H + CH_3.C_6H_4.CH_2,$$

and either its thermal decomposition,

$$CH_3 \cdot C_6H_4 \cdot CH_2 \cdot \;\rightarrow\; H + CH_2{=}\!\!\left\langle\!\!\bigcirc\!\!\right\rangle\!\!{=}CH_2 ,$$

or its disproportionation,

$$2CH_3 \cdot C_6H_4 \cdot CH_2 \cdot \;\rightarrow\; CH_3 \cdot C_6H_4 \cdot CH_3 + CH_2{=}\!\!\left\langle\!\!\bigcirc\!\!\right\rangle\!\!{=}CH_2 ,$$

could form p-xylelene.

The spontaneous polymerization of p-xylelene is initiated by its dimerization,

$$2CH_2{=}\!\!\left\langle\!\!\bigcirc\!\!\right\rangle\!\!{=}CH_2 \;\rightleftarrows\; \cdot CH_2{-}\!\!\left\langle\!\!\bigcirc\!\!\right\rangle\!\!{-}CH_2CH_2{-}\!\!\left\langle\!\!\bigcirc\!\!\right\rangle\!\!{-}CH_2 \cdot ,$$

a reversible exothermic reaction that requires some activation energy. This conclusion follows from some observations to be reported later in the text. At low partial pressure of the monomer and at higher temperatures its equilibrium favors the left side of the equation, and hence the dimerization and the subsequent poly- merization do not take place in the dilute gas phase. The dimeriza- tion may occur in the adsorbed layer provided that the concentration of the monomer is sufficiently high and the temperature sufficiently low. Hence, this film producing reaction occurs only on sufficiently cold surfaces. Thus, objects exposed to the p-xylelene vapor and maintained at room temperature are coated by the polymeric film but its formation is hindered by heating the surface.

The spontaneous formation of a solid film from a gaseous monomer excited my fantasy. However, people knowledgeable in polymer technology told me in those days that an economically viable coating operation would never be carried out in vacuum. It took 20 years until Dr. Gorham working in Union Carbide Corpora- tion demonstrated that such an operation is not only technologically feasible but also economically profitable[9]. Dr. Gorham greatly improved the preparation of p-xylelene by utilizing paracyclophane instead of p-xylene for the pyrolysis yielding the quinonoid hydro- carbon. Moreover, his work and that of his associates revealed the unique features of the vacuum method of coating--a procedure providing advantages unattainable in other coating techniques (see, e.g., ref. 10 for a recent review of this subject).

After discovering p-sylelene and its polymer I searched for other analogous monomers which would be capable of polymerizing spontaneously. Thus I discovered a whole class of similar reactions. For example, pyrolysis of substituted p-xylenes, of 1,4-dimethyl naphthalene or 2,5-dimethyl pyrazine yields the analogues of

p-xylelene that spontaneously polymerize on surfaces exposed to
their vapor. The extension of this work was reported in 1951[11],
and since then, other substituted p-xylelenes were prepared by
Dr. Gorham[9] and by other workers[12] active in this field.

After my move to Syracuse, studies of chemistry of p-xylelene
continued in cooperation with Lou Errede, Bill Landrum, and their
associated in M. W. Kellogg Company. During the following years,
Errede and Landrum[13] succeeded in preparing solutions of p-xylelene
stable at low temperatures by introducing the vapor of this
reactive monomer into vigorously stirred solvents kept at -70°C.
This development led to interesting results. It was shown that the
polymerization of the dissolved p-xylelene ensues whenever a warmer
object, e.g., a glass rod at room temperature, contacts the cold
solution. Apparently the dimerization of the monomer takes place
on the warmer surface and the resulting genuine diradicals propagate
the polymerization. The rate of the propagation increases with
temperature of the solution varied from -60° to -30°C. The ensuing
polymerization exhibits the character of a living one, termination
seems to be prevented presumably because the growing polymers precipitate and thus the active radical ends become "burried" in the
precipitate. Apparently, a similar situation is encountered in the
film formation process. The slow dimerization produces continuously
new growing centers on the surface of a film exposed to the p-xylelene vapor. Their growth leads to the gradual increase in the film's
thickness and burries the centers formed earlier. Thus, free
radicals remain burried in the film and indeed their presence was
demonstrated by the ESR technique[6].

The direct conversion of a gaseous monomer into a solid
crystallinic film, without passing through any intermediate liquid,
prevents flow in the course of coating. Therefore, the thickness
of the coating film remains constant and independent of the curva-
ture of the coated surface, i.e., the film follows truly and
uniformly the shape of the surface. This is one of the outstanding
advantageous features of this coating technique[10]. The thickness
of the film can be easily regulated--it is determined by the time
of exposure of the coated object to the p-xylelene vapor.

Is poly-paraxylelene first formed and then the polymers
crystallize, or does the crystallization proceed simultaneously
with the propagation? This question has been recently investigated
by Wunderlich and his students[14] who studied the behavior of the
p-xylelene-poly-paraxylelene system at various temperatures.

The discovery of poly-paraxylelene led me to ponder more about
various problems of polymer science. I took advantage of my studies
of methyl affinities[15] to determine the reactivities of various
vinyl and diene monomers[16], and strangely enough the studies of
methyl affinities brought me to the discovery of living polymers
and to our extensive investigations of anionic polymerization.

Serendipity played again some role in that development of our research activities. Shortly after publishing the results of our studies of methyl affinities of aromatic hydrocarbons[15a], I met Prof. S. I. Weissman who told me how his series of electron affinities of aromatic hydrocarbons parallels the methyl affinity series. He determined the relative electron affinities by studying electron transfers involving two aromatic hydrocarbons and their radical anions, e.g.,

$$A^{\bar{\cdot}} + B \;\rightleftarrows\; A + B^{\bar{\cdot}}.$$

It occurred to me during this discussion that an unusual situation might be created by transferring an electron to a monomer like styrene. The resulting radical anion could then form two growing centers, by a naively visualized reaction (1)

(1) $\overset{\ominus}{CH}{=}CH_2 + 2CH{=}CH_2 \rightarrow {}^{-}CH \cdot CH_2 - CH \cdot CH_2 - CH_2 \cdot CH \cdot$,
         $\;\;\;\;\;Ph\;\;\;\;\;\;\;\;Ph\;\;\;\;\;\;\;\;\;Ph\;\;\;\;\;\;Ph\;\;\;\;\;\;\;\;\;Ph$

one of them propagating an anionic polymerization while the other initiating a radical polymerization. This idea intrigued me and I asked Weissman whether he tried to transfer electrons to styrene. His reply was straightforward: "No use, it polymerizes." I asked then whether he would mind our looking into this reaction and, having his consent, we started our studies.

Using rigorous vacuum technique and all glass equipment, we mixed sodium naphthalenide in tetrahydrofuran with styrene. We found that the resulting radical anions dimerize forming dianions, viz.,

(2) $2\;\overset{\ominus}{CH}{=}CH_2 \rightarrow {}^{-}CH \cdot CH_2 - CH_2 \cdot CH^{-}$,
         $\;\;\;\;\;Ph\;\;\;\;\;\;\;\;\;\;Ph\;\;\;\;\;\;\;\;\;\;Ph$

and thus two growing centers are formed, but both propagate by anionic mechanism. Moreover, we discovered that the resulting polymerization proceeds without termination and since a terminated polymer, incapable of resuming its growth, is referred to as a dead polymer, I proposed the term "living polymers" for those that retain, virtually indefinitely, their capacity of growth. This turned out to be a successful term rapidly accepted in the international literature.

The lack of termination was demonstrated by a simple experiment[17]. To a solution of sodium naphthalenide, say in 50 ml of tetrahydrofuran, 10 grams of styrene were added. The resulting

fast reaction was over within seconds yielding a viscous red
solution of polystyrene. Its viscosity was measured in situ by
determining the time of falling of a piece of iron enclosed in a
glass envelope. Thereafter, additional 10 g of styrene and 50 ml
of THF were added to the red solution. The added styrene polymerized and the viscosity of the resulting solution greatly
increased. Since the concentration of polystyrene (in grm/ℓ)
remained unaffected by the addition, the increase in viscosity had
to be attributed to the increase of the molecular weight of the
resulting polystyrene. In other words, the previously formed polymers were living and resumed their growth on addition of further
aliquot of styrene.

This proof was followed by another[17]. After polymerizing the
first batch of styrene an aliquot of isoprene was added. After
opening the reaction vessel it was found that both monomers polymerized quantitatively and the resulting polymer was identified
as a block polymer--a chain of styrene units followed by a chain
of isoprene units. The formation of block polymers provides,
therefore, another proof for lack of termination.

The consequences of lack of termination were outlined and
tested. Living polymers are not infinitely long. In any preparation, the amount M of the available monomer is finite and since it
is distributed over all the growing ends, the average degree of
polymerization

$$\overline{DP}_n = M/\underline{I} \text{ or } 2M/\underline{I}$$

depending on whether one or both of the polymer ends are active.
In these equations, $\underline{I}$ denotes the total amount of available
initiator.

Electron-transfer initiation, developed in the course of our
studies, is extremely fast and therefore all the growing centers
are formed simultaneously before they had a chance to grow by more
than a few monomeric units. In a living polymer system termination
or chain transfer is avoided and under such conditions the resulting
macromolecules have the Poisson molecular weight distribution[18].
Such a distribution was eventually achieved, although it was
necessary to recognize first the various disturbing factors discussed elsewhere[19] that had to be eliminated.

Living polymers are not terminated spontaneously; however,
they can be deliberately terminated by adding at a freely chosen
time a reagent converting the active ends into inactive ones.
Such a "killing" procedure allows us to introduce desired end-
groups into the prepared polymer and, if both ends are active,
bifunctional polymers could be synthesized. For example, addition
of CO_2 to living polystyrene initiated by electron transfer yields

vulcanized rubber and shows a considerable tensile strength. However, at higher temperatures it flows and can be molded or extruded. The thermo-plastic rubber, a generic name proposed for these novel materials, astonished many polymer physicists and technologists. Diblock polymers (poly-isoprene)-(poly-styrene) or a terblock (poly-isoprene)-(poly-styrene)-(poly-isoprene) do not show these remarkable properties. Their tensile strength is negligible, even if their molecular weight and composition are identical with those of a terblock polymer having a middle rubbery block and two terminal poly-styrene blocks. The subsequent morphological studies demonstrated that the poly-styrene blocks form small rigid domains in a rubbery matrix and act as giant crosslinking centers. This endows the thermoplastic rubber with its characteristic strength. When the melting point of poly-styrene is reached, the crosslinking action is lost and the material flows then readily.

Milkovich's discovery intensified the research concerned with morphology of block-polymers[23] because the thermoplastic rubber showed promising technological value. Such studies were started even earlier. We worked on the preparation of block-polymers of styrene and ethylene oxide[20] when Paul Rempp visited us. Since these block-polymers revealed some non-conventional properties[20] he repeated in Strasbourg our synthesis and offered these materials to Skulios who showed[24] that they form mesomorphic phases--crystallinic layers of polyethylene oxide separated by amorphous layers of glassy poly-styrene. The formation of regular mesomorphic phases is greatly facilitated by the uniformity of blocks--a characteristic property of block-polymers prepared by living polymer technique. An outstanding example of such a regularity was provided by the work of Galot and Sadron[25] who prepared a variety of block polymers of styrene with butadiene, vinyl pyridine or methyl methacrylate. In their samples one kind of blocks formed matrices through which protruded parallel tubes hexagonally spaced composed of the other blocks.

A spectacular phenomenon was described by another of our students, Ed Vanzo[25]. Uniform block polymers of styrene and butadiene, swollen by a small amount of solvent and spread on the surface, form parallel layers of polystyrene and polybutadiene of uniform thickness. For polymers of a sufficiently high molecular weight the separation of the layers becomes comparable to the wavelength of visible light. Since the uniformly distributed layers act as a grating, the intrinsically colorless material acquires brilliant colors due to the interference of light.

The lack of termination or chain transfer in living polymer systems allows us to study the monomer-polymer equilibria. The principle of microscopic reversibility demands that living polymers possess not only the ability to grow but also the propensity to degrade. Thus, a reversible reaction taking place in the monomer-

α,ω-dicarboxylic acid, reaction with ethylene oxide yields α,ω-
dicarboxylic derivative, etc.[20]   This procedure is of a great
industrial as well as academic value and led to many interesting
products.   Other synthetic methods utilizing living polymers discussed in ref. 21 led to the preparation of tailor-made star-shaped
polymers, comb-shaped polymers, etc.

    The most fruitful synthetic application of living polymers
stemmed from their ability to produce tailor-made block polymers.
For example, we showed in our first papers[17] how the anionic polymerization of styrene followed by the addition of isoprene produced
a block polymer (poly-isoprene) (poly-styrene) (poly-isoprene).
The reverse order of addition gives (poly-styrene) (poly-isoprene)
(poly-styrene).   Since the initiation is rapid and termination
excluded, each block could have a unique size determined easily by
the wish of the operator.   To illustrate the power of this
technique, let us consider the following preparation.   Say that
living poly-A may propagate polymerization of B and vice-versa.
Let us start with one mole of each monomer and initiate the polymerization with a fixed amount of initiator.   If A is added first
and B thereafter, a two-block species is formed,

A........A.B........B.

However, if only one-half of A is first added and thereafter one-
half of B followed by the remaining A and then by the remaining B,
the preparation yields macromolecules of the same size and composition as the previous one but with a different distribution of
monomers along the chain, namely,

A......A.B......B.A......A.B......B.

The generalization of this procedure is obvious, and hence the
above technique permits one to synthesize polymers not only of predetermined size and composition but also of any desired distribution
of units along the chain.

    The advent of living polymer technique revolutionalized the
academic and industrial approach to block-polymers.   It allowed
systematic studies of properties of these materials as functions
of structural factors since the difficulties arising from the
unknown composition, the presence of homo-polymers in the block-
polymers, etc., were avoided in this synthetic method.   A few
examples will illustrate the unexpected achievements.

    A most unusual type of rubber was developed by my former
student, Ralph Milkovich[22], who participated in our first preparation of block-polymers.   Using the living polymer technique, he
synthesized a terblock polymer (poly-styrene)-(poly-isoprene)-(polystyrene).   This polymer behaves at ambient temperature like a

living polymer systems,

$$\text{\textasciitilde\textasciitilde\textasciitilde}M^* + M \underset{\leftarrow}{\rightarrow} \text{\textasciitilde\textasciitilde\textasciitilde}M^*,$$

eventually establishes an equilibrium when $[M] = [M]_e = 1/K$. Here $[M]_e$ denotes the equilibrium concentration of the monomer and K the equilibrium constant of propagation. The equilibrium concentration of the monomer depends on the system, i.e., on the nature of monomer, of solvent and temperature, but its value is not affected by the mechanism of polymerization. Determination of $[M]_e$ over a temperature range allows us to calculate ΔH and ΔS of propagation, and its dependence on polymer concentration and the nature of solvent provides information on the solvent-polymer interaction. A thorough discussion of these topics is reported elsewhere[26]. However, it should be stressed that the above thermodynamic ramifications apply to high-molecular weight polymers and the treatment has to be modified when one deals with oligomers[27].

Application of living polymers to kinetic studies led to the understanding of intricate details of mechanism of ionic polymerization. A solution of living polymers may be prepared at any desired concentration of active ends and then mixed with the respective monomer. Polymerization ensues then and the rate of monomer consumption directly measures the rate of propagation. Since termination is prevented the reaction exhibits a pseudo-first order character[28] because the concentration of growing centers remains constant. From such experiments the values of the pseudo-first order rate constants are calculated and the latter yield a bimolecular rate constant of propagation when plotted versus the concentration of active centers. The latter can be determined before or after completion of the experiment.

Extensive kinetic studies of propagation led to many interesting results. They allowed us to determine not only the rate constants of propagation for a variety of monomers but also, after some modification of the technique, they permitted us to evaluate directly the rate constants, k_{12} and k_{21}, of cross-propagation[29,30] and to assess the effects of penultimate groups upon the rate of copolymerization[31].

Closer examination of the kinetic data revealed intriguing deviations from the simple kinetic scheme outlined above. The polymerization of living polystyrene in dioxane[32] proceeds as expected, the observed bimolecular rate constant being independent of concentration of living polymers, although its value is affected by the nature of the counterions. However, such a polymerization performed in THF behaves differently[33]--the observed bimolecular rate constant of propagation increases with dilution of living polymers. Combined kinetic and conductrometric studies, as well

as studies of the effect exerted by salts sharing a common cation
with the living polymers, conclusively demonstrated the participation of two species in the propagation: free ions and ion pairs.
The former propagate much faster than the latter, and the rate of
their growth is independent of the cation's nature. The participation of free ions and ion pairs in anionic polymerization has been
confirmed for other systems and an extensive literature of this
subject is now available.

Continuation of these studies revealed the complex nature of
ion pairs. Association of a cation with an anion may result in
formation of more than one kind of species. Thus a tight (contact)
or a loose (separated) ion pair may participate in the propagation
process and hence the observed rate constant attributed to ion pair
propagation acquires the form

$$\underline{k}_{p,\text{ion pairs}} = \underline{f}\underline{k}_t + (1-\underline{f})\underline{k}_\ell.$$

Here $\underline{k}_t$ and $\underline{k}_\ell$ refer to the propagation constants of tight and
loose ion pairs, respectively, while $\underline{f}$ denotes the fraction of the
tight pairs present in the equilibrated system. In contrast to
the system, free ions and ion pairs, where the fraction of free ions
varies with the total concentration of living species, the fraction
$\underline{f}$ is concentration independent. However, since $\underline{f}$ increases with
temperature the proportion of loose pairs progressively becomes
higher as the temperature of the solution decreases. In most
systems, although not in all, the loose pairs are more reactive
than the tight ones, and therefore the rate of propagation caused
by ion pairs may increase with decreasing temperature. Such a
"negative" activation energy of propagation was observed in the
system sodium-polystyryl-THF[34]. However, further decrease of the
temperature must lead eventually to a decrease in rate of polymerization, because for $\underline{f}$ sufficiently small and $(1-\underline{f})$ approaching
unity subsequent temperature decrease virtually does not increase
the concentration of the reactive species. At this temperature
range the genuine activation energy of the loose pairs' propagation
governs the temperature dependence of the polymerization. Such a
phenomenon was observed in the system sodium polystyryl-DME[34]. As
the temperature decreases the pertinent rate constant initially
increases, reaches a maximum, and then decreases.

Tight ion pairs present in poorly solvated solvents may be
converted into much more reactive loose pairs by addition to their
solution of small amounts of powerfully solvating agents. The
dramatic effects arising from addition of tetraglyme to the
solution of sodium polystyryl in tetrahydropyrane were reported by
Shinohara, Smid, and Szwarc[35]. The rate of propagation increased
100-fold on the addition of 2.8×10^{-3} M glyme.

The rate of polymerization may be also retarded by the addition of compounds converting living polymers into dormant ones. For example, addition of anthracene (A) to living polystyrene ($\sim\sim\sim S^-$) converts the latter into a dormant adduct[36]

$$\sim\sim\sim S^- + A \; \underset{\leftarrow}{\rightarrow} \; \sim\sim\sim SA^-.$$

The association is reversible, hence the concentration of the un-associated living polystyrene decreases as the concentration of anthracene increases. Thus, the rate of polymerization may be reduced at wish by the addition of an appropriate amount of anthracene.

An interesting example of intramolecular conversion of living polymer into a dormant one was observed on addition of styrene to living poly-1-vinyl naphthalene[37]. The first molecule of styrene added to the terminal $-\overline{C}H$(Naphthyl) anion yields a benzyl carbanion intramolecularly associated with the preceding naphthyl group. Such an anion is dormant, although it remains in equilibrium with a small fraction of unassociated $\sim\sim\sim\overline{C}H$(Ph) groups. The addition of a second styrene unit is therefore very slow; however, after it does take place it forms a $\sim\sim\sim CH_2CH(Ph) \cdot CH_2\overline{C}H(Ph)$ group non-associated with naphthyl group because the two groups are now separated. Consequently, the third, fourth, etc., styrene units are added rapidly in spite of the exceedingly slow addition of the second group. The kinetics of this reaction was unravelled and the rate was shown to increase with time.

The nature of the cation-living anion pair may be modified by intramolecular interaction. An example is provided by the salts of living poly-2-vinyl-pyridine and poly-vinyl-quinoline[38,39]. The association with the nitrogen of the heteromoiety attached to the last segment of the living polymer converts this intrinsically tight pair into a loose one. As a result, these ion-pairs are much more reactive than anticipated even in solvents of low solvating power, such as dioxane and tetrahydropyrane.

Anionic polymerization may be also affected by the formation of higher ionic aggregates. For example, cesium salt of living polystyrene endowed with two active end groups dissociates into

$$Cs^+, {}^-S\sim\sim\sim S^-, Cs^+ \; \underset{\leftarrow}{\rightarrow} \; Cs^+, {}^-S\sim\sim\sim S^- + Cs^+.$$

The partially dissociated species undergoes an intramolecular association leading to a cyclic triple ion[40]

$$Cs^+, {}^-S\sim\sim\sim S^- \; \underset{\leftarrow}{\rightarrow} \; \overparen{S^-, Cs^+, S^-}$$

Thus the rapidly growing ~~~~~S^- ions are removed from the system
and their replenishment through the previously mentioned dissociation increases the concentration of the free Cs^+ ions and thus
decreases the equilibrium concentration of the free ~~~~~S^- ions.
In result, the rate of polymerization becomes slower than expected
while the conductance of the solution becomes higher.

A surprising phenomenon caused by the formation of higher
ionic aggregates was discovered during our study of polymerization
of barium salt of living polystyrene[41] in THF. In this system
the pseudo-first order rate constant of propagation was unaffected
by the dilution of living polymers, although the investigated salt
dissociated into conducting species. It was found that the conductance mainly arises from the formation of

$$(\text{~~~~~}S^-, Ba^{2+}) \text{ cations and } (\text{~~~~~}S^-)_3 Ba^{2+} \text{ anions, viz.,}$$

$$2(\text{~~~~~}S^-)_2 Ba^{2+} \rightleftarrows (\text{~~~~~}S^-, Ba^{2+})^+ + \{(\text{~~~~~}S^-)_3, Ba^{2+}\}^-,$$

because the dissociation

$$(\text{~~~~~}S^-)_2 Ba^{2+} \rightleftarrows (\text{~~~~~}S^-, Ba^{2+})^+ + \text{~~~~~}S^-$$

is exceedingly low. Nevertheless, the rapidly growing ~~~~~S^-
free ions are produced by the latter dissociation, the contribution
of

$$(\text{~~~~~}S^-)_2 Ba^{2+}, \ (\text{~~~~~}S^-, Ba^{2+}),$$

and $(\text{~~~~~}S^-)_3 Ba^{2+}$ to the polymerization being negligible. Inspection of the above equilibria shows that the ratio

$$(\text{~~~~~}S^-, Ba^{2+})/(\text{total salt})$$

is independent of the salt's concentration (the degree of dissociation is minute), and hence the concentration of the growing ~~~~~S^-
ions, given by const.x(total salt)/(~~~~~S^-, Ba^{2+}), is also independent
of the salt's concentration. This accounts for the constant rate of
polymerization at a given monomer concentration, independent of
concentration of living polymers.

Still more complex associations of living polymers are observed
in systems involving lithium counterions and hydrocarbon solvents.
For example, in benzene lithium polystyrene is dimeric (~~~~~$S^-, Li^+)_2$
and the polymerization is propagated by minute amount of non-associated ~~~~~S^-, Li^+ which remains in equilibrium with the dimer. The
behavior of these systems is discussed in ref. 42.

Although studies of anionic polymerization led to the discovery
of living polymers, their existence need not be restricted to

anionic systems only. In 1963, I was appointed Visiting
Professor in the University of Liverpool. There, I met Peter and
Pat Dreyfuss who were looking for an interesting research project.
I suggested investigating the cationic polymerization of tetrahydrofuran initiated by PF_5 and to look into whether the system
approaches a polymer-monomer equilibrium. As pointed out previously,
in a living polymer system the equilibrium concentration of monomer
should be independent of the concentration of an initiator but it
should be affected by temperature. This indeed turned out to be
the case[43], and thus, the extensive studies of living, cationically
propagated polymers was initiated[44].

Like in anionic systems, living, cationically growing polymers
facilitated kinetic studies of these polymerizations. For example,
Stan Penczek, who spent one year with us, demonstrated after his
return to Poland that free ions and ion pairs participate in the
cationic polymerization of tetrahydrofuran and succeeded to
determine the individual rate constants of these species[45]. However, in this system still another species, not yet encountered
in anionic polymerization, propagates the polymerization, namely,
the covalently bonded ester.

The participation of free ions and a variety of ion pairs
induced us to a more general study of the influence of ion pairing
on the reactivity of organic ions and radical ions. We studied
the effect of pairing on the rates of electron-transfer, proton-
transfer, dimerization, etc. Moreover, we attempted to determine
the rates of conversion of tight pairs into loose ones and vice-
versa. A general survey of these topics is reported in the two
volumes of Ions and Ion Pairs in Organic Reactions edited by me[46].

Protonation of vinyl monomers initiates their cationic
polymerization. Hopefully, the presence of the resulting carbonium
ions might be observed spectrophotometrically and then their concentration in the polymerizing system could be determined. In
cooperation with Prof. David Pepper, who stayed with us for 6
months, we investigated by stop-flow technique polymerization of
styrene initiated by perchloric acid[47]. The formation of the
$-CH_2^+CH(Ph)$ was indeed observed and the propagation constant calculated. This study is continued by Pepper and the detailed
features of the reaction require further clarification.

Successful elucidation of the mechanism of electron-transfer
initiation spurred us to a broader study of electron-transfer
reactions. We investigated the dimerization of radical anions
leading to dimeric dianions[48] as well as the reverse processes
resulting in their decomposition[19], and more generally we
intensively studied the chemistry of radical anions. Interestingly
these investigations, not concerned with subjects pertinent to
polymer chemistry or polymer physics, allowed us to solve an

intriguing problem of polymer physics. Chains of polymers are flexible and these molecules continually change their conformation when dissolved in solvents. During these Brownian motions, their two ends approach occasionally each other, then become separated, come closer again, etc. How frequently do these intramolecular "collisions" occur? How does this frequency depend on their length, their nature, the nature of solvent, and the temperature of solution? We developed a relatively simple method to answer these questions. We synthesized chains such as $-(CH_2)_n-$ or $-(CH_2CH_2O_m-CH_2CH_2)-$ terminated by aromatic moieties one of which, but not the other, endowed with an extra-electron. Thus, these molecules become paramagnetic and show their characteristic ESR spectra. In dilute solutions, the shape of such a spectrum is modified by the frequency of intramolecular electron-transfer-- the aromatic end group endowed with an extra electron donating it to the other end group. The transfer results from an intramolecular "collision" and if it is a diffusion controlled process the frequency of transfer measures the frequency of collisions. The latter depends on the dynamic stiffness of the chain and thus such studies may lead to a scale of intrinsic dynamic stiffnesses of polymeric chains. This approach turned out to be successful and the results of these investigations have been reported[50].

Finally, I wish to thank the National Science Foundation which has continually supported our investigations since my arrival to Syracuse in 1953.

REFERENCES

1. M. Szwarc, *J. Chem. Phys.*, <u>16</u>, 128 (1948).
2. M. Szwarc, *Chem. Rev.*, <u>47</u>, 75 (1950).
3. M. Szwarc and J. S. Roberts, *J. Chem. Phys.*, <u>16</u>, 609 (1948).
4. J. S. Roberts and M. Szwarc, *J. Chem. Phys.*, <u>16</u>, 981 (1948).
5. M. Szwarc, *Disc. Faraday Soc.*, <u>2</u>, 39 (1947).
6. (a) A. Namiott, M. Diatkina, and J. Syrkin, *Compt rend. Acad. Sci. USSR*, <u>48</u>, 285 (1945).
    (b) M. Diatkina and J. Syrkin, *Acta Physiochem USSR*, <u>21</u>, 23 (1946).
    (c) C. A. Coulson, D. P. Craig, A. Maccoll, and A. Pullman, *Disc. Faraday Soc.*, <u>2</u>, 36 (1947).
7. C. J. Brown and A. C. Farthing, *Nature*, <u>164</u>, 915 (1949).
8. A. C. Farthing, *J. Chem. Soc.*, (London) 3261 (1953).
9. (a) W. H. Gorham, *J. Polymer Sci.*, <u>4</u>, A1, 3027 (1966).
    (b) W. H. Gorham, *Adv. Chem. Series*, <u>91</u>, 641 (1969).
10. M. Szwarc, *Polymer Science and Engineering*.
11. M. Szwarc, *J. Polymer Sci.*, <u>6</u>, 319 (1951).
12. J. M. Pearson, H. D. Lix, D. J. Williams and M. Levy, *J. Am. Chem. Soc.*, <u>93</u>, 5034 (1971).
13. L. A. Errede and B. F. Landrum, *J. Am. Chem. Soc.*, <u>79</u>, 4952 (1957).

14. (a) S. Kubo and B. Wunderlich, *J. Appl. Physics*, <u>42</u>, 4558, 4565 (1971).
    (b) S. Kubo and B. Wunderlich, *Makromol. Chem.*, 162, 1 (1972).
    (c) G. Treiber, K. Bohlke, A. Wertz, and B. Wunderlich, *J. Poly. Sci.*, <u>11</u>, 1111 (1973).

15. (a) M. Levy and M. Szwarc, *J. Chem. Phys.*, <u>22</u>, 1621 (1954).
    (b) J. H. Binks and M. Szwarc, *Proc. Chem. Soc.*, London, p. 226 (1958).

16. (a) F. Leavitt, M. Levy, M. Szwarc, and V. T. Stannett, *J. Am. Chem. Soc.*, <u>77</u>, 5493 (1955).
    (b) R. P. Buckley and M. Szwarc, *J. Am. Chem. Soc.*, <u>78</u>, 5696 (1956).
    (c) R. P. Buckley, A. Rembaum, and M. Szwarc, *J. Poly. Sci.*, <u>24</u>, 135 (1957).
    (d) A. Rajbenbach and M. Szwarc, *J. Am. Chem. Soc.*, <u>79</u>, 6343 (1957).

17. (a) M. Szwarc, *Nature*, <u>178</u>, 1168 (1956).
    (b) M. Szwarc, M. Levy, and R. Milkovich, *J. Am. Chem. Soc.*, <u>78</u>, 2656 (1956).

18. P. J. Flory, *J. Am. Chem. Soc.*, <u>62</u>, 1561 (1940).

19. M. Szwarc, <u>Carbanions, Living Polymers and Electron Transfer Reactions</u>, pp. 27-69, John Wiley, Pub. (1968).

20. D. H. Richards and M. Szwarc, *Trans. Faraday Soc.*, <u>55</u>, 1644 (1959).

21. Ref. 19, pp. 90-98.

22. (a) G. Holden and R. Milkovich, U. S. Patent 3,265,765 (1966).
    (b) G. Holden, E. T. Bishop, and N. R. Legge, *J. Poly. Sci.*, Part C, <u>26</u>, 37 (1969).

23. <u>Block and Graft Copolymers</u>, J. J. Burke and V. Weiss, eds., Syracuse University Press (1973).

24. (a) A. Skulios, G. Finaz, and J. Parrod, *Compt. Rend.*, <u>251</u>, 739 (1960).
    (b) A. Skulios and G. Finaz, *ibid*, <u>252</u>, 3467 (1961).

25. E. Vanzo, *J. Poly. Sci.*, A-1, <u>4</u>, 1727 (1966).

26. Ref. 19, pp. 104-148.

27. (a) A. Vrancken, J. Smid, and M. Szwarc, *J. Am. Chem. Soc.*, <u>83</u>, 2772 (1961) and *Trans. Faraday Soc.*, <u>58</u>, 2036 (1962).
    (b) C. L. Lee, J. Smid, and M. Szwarc, *J. Am. Chem. Soc.*, <u>85</u>, 912 (1963).

28. C. Geacintov, J. Smid, and M. Szwarc, *J. Am. Chem. Soc.*, <u>84</u>, 2508 (1962).

29. (a) C. L. Lee, J. Smid, and M. Szwarc, *J. Am. Chem. Soc.*, <u>83</u>, 2961 (1961).
    (b) D. N. Bhattacharyya, C. L. Lee, J. Smid, and M. Szwarc, *ibid*, <u>85</u>, 533 (1963).

30. M. Shima, D. N. Bhattacharyya, J. Smid, and M. Szwarc, *J. Am. Chem. Soc.*, <u>85</u>, 1306 (1963).

31. C. L. Lee, J. Smid, and M. Szwarc, *Trans. Faraday Soc.*, <u>59</u>, 1192 (1963).

32. D. N. Bhattacharyya, J. Smid, and M. Szwarc, *J. Phys. Chem.*, <u>69</u>, 624 (1965).

33. D. N. Bhattacharyya, C. L. Lee, J. Smid, and M. Szwarc, *Polymer*, <u>5</u>, 54 (1964), *J. Phys. Chem.*, <u>69</u>, 612 (1965).

34. T. Shimomura, K. J. Tölle, J. Smid, and M. Szwarc, *J. Am. Chem. Soc.*, <u>89</u>, 796 (1967).

35. M. Shinohara, J. Smid, and M. Szwarc, *J. Am. Chem. Soc.*, <u>90</u>, 2175 (1968), *Chem. Comm.*, London, 1232 (1969).

36. (a) S. N. Khanna, M. Levy, and M. Szwarc, *Trans. Faraday Soc.*, <u>58</u>, 747 (1962).
    (b) R. Asami, S. N. Khanna, M. Levy, and M. Szwarc, *Trans. Faraday Soc.*, <u>58</u>, 1821 (1962).

37. F. Bahsteter, J. Smid, and M. Szwarc, *J. Am. Chem. Soc.*, <u>85</u>, 3909 (1963).

38. (a) M. Fisher and M. Szwarc, *Makromolec.*, <u>3</u>, 23 (1970).
    (b) A. Rigo, M. Szwarc, and G. Sackmann, *ibid.*, <u>4</u>, 622 (1971).

39. M. Tardi, D. Rouge and P. Sigwalt, *Europ. Polymer J.*, <u>3</u>, 85 (1967).

40. D. N. Bhattacharyya, J. Smid, and M. Szwarc, *J. Am. Chem. Soc.*, <u>86</u>, 5024 (1964).

41. M. van Beylen, B. De Groot, and M. Szwarc, *Macromolec.*, <u>8</u>, 396 (1975).

42. Ref. 19, pp. 476-519.

43. P. and M. P. Dreyfuss, *Polymer*, (London).

44. P. and M. P. Dreyfuss, in <u>Ring Opening Polymerization</u>, K. C. Frisch and S. L. Reegen, eds., Marcel Decker Publ. (1969).

45. S. Penczek, in Symposium on Cationic Polymerization, Akron, 1976.

46. <u>Ions and Ion Pairs in Organic Reactions</u>, M. Szwarc, ed., Volume I (1972), Volume II (1974), John Wiley Pub.

47. Michael de Sorgo, David C. Pepper, and Michael Szwarc, *J.C.S. Chem. Comm.*, 419 (1973).

48. (a) C. L. Lee, J. Smid, and M. Szwarc, *J. Phys. Chem.*, <u>66</u>, 904 (1962).
    (b) G. Spach, H. Monteiro, M. Levy, and M. Szwarc, *Trans. Faraday Soc.*, <u>58</u>, 1809 (1962).
    (c) J. Jagur, M. Levy, M. Feld, and M. Szwarc, *Trans. Faraday Soc.*, <u>58</u>, 2168 (1962).
    (d) M. Szwarc, presented as a Main Lecture in the Annual Meeting of the Bunsen Gesellschaft, Mainz, Germany, May 1963. Ber. Bunsen Gesell. <u>67</u>, 763 (1963).
    (e) M. Matsuda, J. Jagur-Grodzinski, and M. Szwarc, *Proc. Roy. Soc.*, A <u>288</u>, 212 (1965).

49. (a) M. Szwarc and R. Asami, *J. Am. Chem. Soc.*, <u>84</u>, 2269 (1962).
    (b) J. Jagur-Grodzinski and M. Szwarc, *Trans. Faraday Soc.*, <u>59</u>, 2305 (1963).
    (c) D. Gill, J. Jagur-Grodzinski, and M. Szwarc, *Trans. Faraday Soc.*, <u>60</u>, 1424 (1964).

50. (a) H. D. Connor, K. Shimada, and M. Szwarc, *Macromolecules*, <u>5</u>, 801 (1972).

(b) Michael Szwarc, in 22nd Nobel Symposium on "ESR
Applications to Polymer Research," p. 291 (1973).
(c) K. Shimada and M. Szwarc, *J. Am. Chem. Soc.*, 97, 3313
(1975).
(d) K. Shimada and M. Szwarc, *J. Am. Chem. Soc.*, 97, 3321
(1975).
(e) K. Shimada, Y. Shimozato, and M. Szwarc, *J. Am. Chem. Soc.*,
97, 5834 (1975).

Index